Mechanics of Fluids

Solutions Manual

Mechanics of Fluids

Eighth edition

Solutions manual

Bernard Massey
Reader Emeritus in Mechanical Engineering
University College, London

Revised by

John Ward-Smith
Formerly Senior Lecturer in Mechanical Engineering
Brunel University

Taylor & Francis
Taylor & Francis Group
LONDON AND NEW YORK

Seventh edition published by Stanley Thornes (Publishers) Ltd in 1998

Eighth edition published 2006
by Taylor & Francis
2 Park Square, Milton Park, Abingdon, Oxon OX14 4RN

Simultaneously published in the USA and Canada
by Taylor & Francis
270 Madison Ave, New York, NY 10016

Taylor & Francis is an imprint of the Taylor & Francis Group

Typeset in Sabon by
Newgen Imaging Systems (P) Ltd, Chennai, India
Printed and bound in Great Britain by
Bell & Bain Ltd, Glasgow

British Library Cataloguing in Publication Data
A catalogue record for this book is available from the British Library

Library of Congress Cataloging in Publication Data
A catalog record for this book has been requested

ISBN 0–415–36204–0

Chapter 1

1.1 Since $pV = mRT$, $\dfrac{V_1}{V_1} = \dfrac{T_1 p_2}{T_2 p_1}$

$$\therefore V_1 = \frac{\pi}{6}(20\ \text{m})^3 \frac{288.15}{233.15}\frac{1.1}{101.3} = \mathbf{56.2\ m^3}$$

1.2
$$\varrho = \frac{p}{RT} = \frac{1.4 \times 10^5\ \text{N}\cdot\text{m}^{-2}}{287\ \text{J}\cdot\text{kg}^{-1}\cdot\text{K}^{-1} \times 323.15\ \text{K}} = \mathbf{1.51\ kg\cdot m^{-3}}$$

1.3 $K = \varrho \dfrac{\partial p}{\partial \varrho}$ Assume K constant. Then $\ln(\varrho/\varrho_0) = \dfrac{p - p_0}{K}$

$$\therefore \varrho = \varrho_0 \exp\left(\frac{p - p_0}{K}\right) = 1025\ \text{kg}\cdot\text{m}^{-3} \exp\left(\frac{81.7 \times 10^6}{2.34 \times 10^9}\right)$$

$$= \mathbf{1061\ kg\cdot m^{-3}}$$

1.4
$$\varrho = \frac{\mu}{\nu} = \frac{2 \times 10^{-5}\ \text{N}\cdot\text{s}\cdot\text{m}^{-2}}{15 \times 10^{-6}\ \text{m}^2\cdot\text{s}^{-1}} = 1.333\ \text{kg}\cdot\text{m}^{-3}$$

$$R = \frac{p}{\varrho T} = \frac{1.013 \times 10^5\ \text{N}\cdot\text{m}^{-2}}{1.333\ \text{kg}\cdot\text{m}^{-3} \times 293.15\ \text{K}} = 259.2\ \text{J}\cdot\text{kg}^{-1}\cdot\text{K}^{-1}$$

$$\therefore M = \frac{8310}{259.2} = \mathbf{32.06}$$

1.5 $\mu = \nu\varrho = 400 \times 10^{-6}\ \text{m}^2\cdot\text{s}^{-1} \times 850\ \text{kg}\cdot\text{m}^{-3} = 0.34\ \text{Pa}\cdot\text{s}$

$$\text{Velocity gradient} = \frac{0.12\ \text{m}\cdot\text{s}^{-1}}{0.1 \times 10^{-3}\ \text{m}} = 1200\ \text{s}^{-1}$$

$$\text{Area} = \pi 0.2 \times 1.2\ \text{m}^2 = 0.754\ \text{m}^2$$

$$\text{Force} = 0.754\ \text{m}^2 \times 0.34\ \text{Pa}\cdot\text{s} \times 1200\ \text{s}^{-1} = \mathbf{307.6\ N}$$

1.6 $$\text{Total force on plate} = \text{Area} \times \mu \left\{ \left(\frac{\partial u}{\partial y}\right)_{\text{side A}} + \left(\frac{\partial u}{\partial y}\right)_{\text{side B}} \right\}$$

$$= (0.25\ \text{m})^2 \times 0.7\ \text{Pa}\cdot\text{s} \left\{ \frac{0.15\ \text{m}\cdot\text{s}^{-1}}{0.006\ \text{m}} + \frac{0.15\ \text{m}\cdot\text{s}^{-1}}{0.019\ \text{m}} \right\}$$

$$= \mathbf{1.439\ N}$$

1.7 For annulus, radius r, width δr

$$\text{Force} = \text{Area} \times \mu \times \frac{\text{Velocity}}{\text{Clearance}} = 2\pi r \delta r \mu \frac{\omega r}{c}$$

$$\therefore \text{Torque} = \text{Force} \times r = 2\pi r^3 \delta r \frac{\mu\omega}{c}$$

$$\text{Total torque} = \int_0^R 2\pi r^3 \frac{\mu\omega}{c} \mathrm{d}r = \frac{\pi R^4 \mu\omega}{2c}$$

$$= \frac{\pi (0.1\ \text{m})^4 0.14\ \text{Pa}\cdot\text{s} \times 2\pi \times 7\ \text{rad}\cdot\text{s}^{-1}}{2 \times 0.00013\ \text{m}} = \mathbf{7.44\ N\cdot m}$$

1.8 $$p = \frac{2\gamma}{d} = \frac{2 \times 0.073\ \text{N}\cdot\text{m}^{-1}}{0.004\ \text{m}} = \mathbf{36.5\ Pa}$$

1.9 $$h = \frac{4\gamma \cos\theta}{\varrho g d} = \frac{4 \times 0.073\ \text{N}\cdot\text{m}^{-1} \times 1}{1000\ \text{kg}\cdot\text{m}^{-3} \times 9.81\ \text{N}\cdot\text{kg}^{-1} \times 0.005\ \text{m}}$$

$$= 0.00595\ \text{m} = \mathbf{5.95\ mm}$$

1.10 $$h = \frac{4 \times 0.377\ \text{N}\cdot\text{m}^{-1} \times \cos 140^\circ}{(13.56 - 1)1000\ \text{kg}\cdot\text{m}^{-3} \times 9.81\ \text{N}\cdot\text{kg}^{-1} \times 0.006\ \text{m}}$$

$$= \mathbf{-1.563\ mm}$$

1.11 $$Re = \frac{ud\varrho}{\mu} = \frac{4Q\varrho}{\pi d\mu} = \frac{4 \times 0.0025\ \text{m}^3\cdot\text{s}^{-1} \times 900\ \text{kg}\cdot\text{m}^{-3}}{\pi 0.05\ \text{m} \times 0.038\ \text{N}\cdot\text{s}\cdot\text{m}^{-2}} = \mathbf{1508}$$

$$u = \frac{2000\mu}{d\varrho} = \frac{2000 \times 0.038\ \text{N}\cdot\text{s}\cdot\text{m}^{-2}}{0.05\ \text{m} \times 900\ \text{kg}\cdot\text{m}^{-3}} = \mathbf{1.689\ m\cdot s^{-1}}$$

1.12 $$Re = \frac{4Q\varrho}{\pi d\mu} = \frac{4 \times 0.01\ \text{m}^3\cdot\text{s}^{-1}}{\pi 0.08\ \text{m} \times 370 \times 10^{-6}\ \text{m}^2\cdot\text{s}^{-1}} = \mathbf{430}\ \therefore \mathbf{Laminar}$$

Chapter 2

2.1 $h = \dfrac{p}{\varrho g} = \dfrac{200 \times 10^3\ \text{N}\cdot\text{m}^{-2}}{1590\ \text{kg}\cdot\text{m}^{-3} \times 9.81\ \text{N}\cdot\text{kg}^{-1}} = \mathbf{12.82\ m}$

2.2 Pressure depends only on depth below free surface.

(a) $p = \varrho g h = (820\ \text{kg}\cdot\text{m}^{-3} \times 9.81\ \text{N}\cdot\text{kg}^{-1})(3 - 0.15)\ \text{m}$

$= 22\,930\ \text{N}\cdot\text{m}^{-2} = \mathbf{22.93\ kPa}$

(b) $p = 820 \times 9.81\ \text{N}\cdot\text{m}^{-3} \times (3 + 2)\ \text{m} = \mathbf{40.2\ kPa}$

(c) $p = 820 \times 9.81\ \text{N}\cdot\text{m}^{-3} \times \{3 + 2 - (1.2 \sin 30^\circ + 0.6)\}\ \text{m}$

$= 820 \times 9.81 \times 3.8\ \text{N}\cdot\text{m}\ ^{2} = \mathbf{30.57\ kPa}$

(d) Load = Pressure × Area

$= 820 \times 9.81 \times 3\ \text{N}\cdot\text{m}^{-2} \times (3.5 \times 2.5)\ \text{m}^2 = \mathbf{211.2\ kN}$

2.3 $h_{\text{air}} = \dfrac{p}{\varrho_{\text{air}} g} = \dfrac{\varrho_{\text{water}} g h_{\text{water}}}{\varrho_{\text{air}} g} = \dfrac{\varrho_{\text{water}}}{\varrho_{\text{air}}} h_{\text{water}}$

$= \dfrac{1000\ \text{kg}\cdot\text{m}^{-3} \times 287\ \text{J}\cdot\text{kg}^{-1}\cdot\text{K}^{-1} \times 288.15\ \text{K}}{1.013 \times 10^5\ \text{N}\cdot\text{m}^{-2}} 0.075\ \text{m}$

$= \mathbf{61.2\ m}$

2.4 $pV = \text{constant}$

$\therefore \left(\dfrac{d}{4\ \text{mm}}\right)^3 = \dfrac{101.3 \times 10^3\ \text{Pa} + 1000\ \text{kg}\cdot\text{m}^{-3} \times 9.81\ \text{N}\cdot\text{kg}^{-1} \times 9\ \text{m}}{101.3 \times 10^3\ \text{Pa}}$

whence $d = \mathbf{4.93\ mm}$

2.5 $\Delta p = 820\ \text{kg}\cdot\text{m}^{-3} \times 9.81\ \text{N}\cdot\text{kg}^{-1} \times 2\ \text{m} + (13.56 - 0.82)$

$\times 1000\ \text{kg}\cdot\text{m}^{-3} \times 9.81\ \text{N}\cdot\text{kg}^{-1} \times 0.225\ \text{m} = \mathbf{44.2\ kPa}$

$$\frac{\Delta p^*}{\varrho g} = \Delta h = \frac{0.225\ \text{m}(13.56 - 0.82)1000\ \text{kg}\cdot\text{m}^{-3} \times 9.81\ \text{N}\cdot\text{kg}^{-1}}{820\ \text{kg}\cdot\text{m}^{-3} \times 9.81\ \text{N}\cdot\text{kg}^{-1}}$$

$$= \mathbf{3.496\ m}$$

$$44\,200\ \text{N}\cdot\text{m}^{-2} = 820 \times 9.81 \times 2\ \text{N}\cdot\text{m}^{-2} + x(0.82 - 0.74)1000 \times 9.81\ \text{N}\cdot\text{m}^{-3}$$

whence $x = \mathbf{35.83\ m}$

2.6

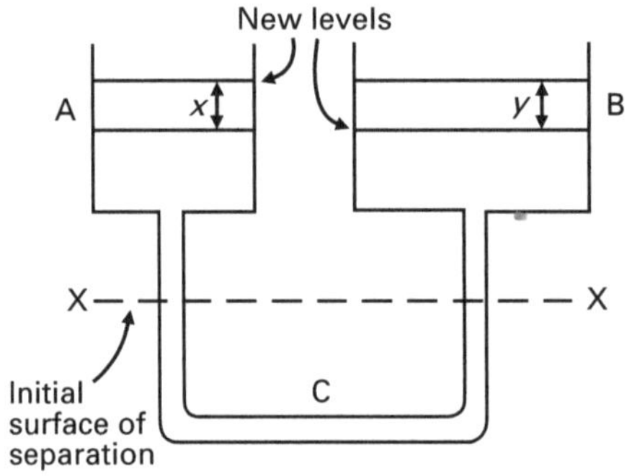

Movement of fluid in

$$\text{C} = 60\ \text{mm} \times 70\ \text{mm}^2 = (500\ \text{mm}^2)x = (800\ \text{mm}^2)y$$

$$\therefore x = 8.4\ \text{mm};\quad y = 5.25\ \text{mm}$$

Measuring above XX: Initially $0.8h_A = 0.9h_B$

Later: $800 \times 9.81(\text{Old } h_A - 60 + 8.4)10^{-3}\ \text{Pa}$

$$= p + 900 \times 9.81(\text{Old } h_B - 60 + 5.25)10^{-3}\ \text{Pa}$$

$$\therefore\ p = 9.81 \times 10^{-3}(-800 \times 51.6 + 900 \times 65.25)\text{Pa} = \mathbf{171.1\ Pa}$$

2.7 From eqn 2.7 $p = p_0\left(1 - \frac{\lambda z}{T_0}\right)^{g/R\lambda}$

$$= 101.5\ \text{Pa}\left(1 - \frac{0.0065 \times 7500}{288.15}\right)^{9.81/287\times 0.0065}$$

$$= \mathbf{38.3\ kPa}$$

2.8 $$\frac{p}{p_0} = \left(\frac{T_0 - \lambda z}{T_0}\right)^{g/R\lambda} = \left(\frac{T_{top}}{T_{top} + \lambda z}\right)^{g/R\lambda}$$

$$\therefore z = \frac{T_{top}}{\lambda}\left\{\left(\frac{p_0}{p}\right)^{R\lambda/g} - 1\right\}$$

$$= \frac{268.15}{0.0065}\ \text{m}\left\{\left(\frac{749}{566}\right)^{287\times 0.0065/9.81} - 1\right\}$$

$$= \mathbf{2257\ m}$$

2.9 $F = (1.2 \times 1.8)\ \text{m}^2 \times 1000\ \text{kg}\cdot\text{m}^{-3} \times 9.81\ \text{N}\cdot\text{kg}^{-1} \times (x + 0.9 \sin 30°)\ \text{m}$

(a) $2160 \times 9.81\ \text{N}\cdot\text{m}^{-1} \times 0.45\ \text{m} = \mathbf{9.54\ kN}$
(b) $2160 \times 9.81\ \text{N}\cdot\text{m}^{-1} \times 0.95\ \text{m} = \mathbf{20.13\ kN}$
(c) $2160 \times 9.81\ \text{N}\cdot\text{m}^{-1} \times 30.45\ \text{m} = \mathbf{645\ kN}$

Centre of pressure is at slant depth $\dfrac{(bd^3/12) + bd(2x+0.9)^2}{bd(2x+0.9)}$

$$= \frac{d^2}{12(2x+0.9)} + 2x + 0.9(\text{metres})$$

$$= \frac{(1.8\ \text{m})^2}{12(2x+0.9)} + 2x + 0.9\ \text{m}$$

$$\text{that is } \left\{\frac{1.8^2}{12(2x+0.9)} + 0.9\right\} \text{m from upper edge}$$

= (a) **1.2 m**; (b) **1.042 m**; (c) **0.904 m** from upper edge

2.10 By symmetry, centre of pressure is on vertical centre-line

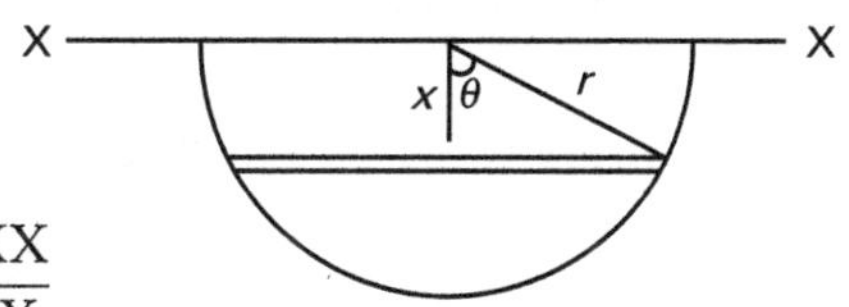

$$\text{Depth} = \frac{\text{2nd moment about XX}}{\text{1st moment about XX}}$$

$$= \frac{\int_0^r x^2 2(r^2 - x^2)^{1/2} dx}{\int_0^r x 2(r^2 - x^2)^{1/2} dx}$$

$$= \frac{\int_{\pi/2}^0 (r\cos\theta)^2 2r\sin\theta(-r\sin\theta d\theta)}{\int_{\pi/2}^0 r\cos\theta 2r\sin\theta(-r\sin\theta d\theta)}$$

$$= \frac{r\int_0^{\pi/2} \cos^2\theta \sin^2\theta d\theta}{\int_0^{\pi/2} \sin^2\theta \cos\theta d\theta}$$

$$= \frac{r\int_0^{\pi} \frac{1}{8}\sin^2 2\theta d(2\theta)}{\left[\frac{1}{3}\sin^3\theta\right]_0^{\pi/2}}$$

$$= \frac{r/8\,[2\theta/2 - (1/4)\sin 4\theta]_0^{2\theta=\pi}}{1/3}$$

$$= \frac{3}{8} r \frac{\pi}{2} = \mathbf{\frac{3\pi d}{32}}$$

2.11 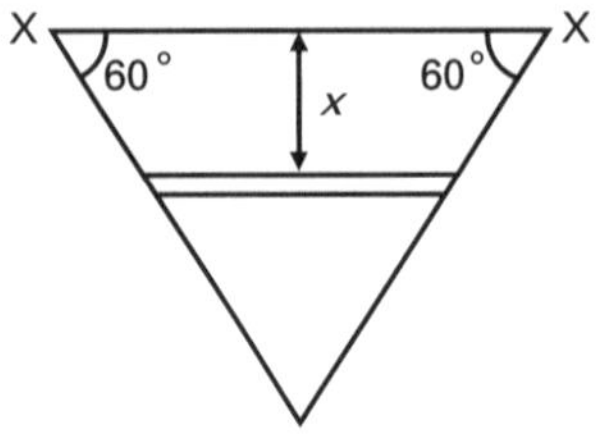

Full depth $= (2.5\text{ m})\sin 60°$

Breadth of strip

$$= 2.5\text{ m}\left\{\frac{(2.5\text{ m})\sin 60° - x}{(2.5\text{ m})\sin 60°}\right\}$$

$$= 2.5\text{ m} - x\operatorname{cosec} 60°$$

∴ Second moment of area about XX

$$= \int_0^{(2.5\text{ m})\sin 60°} (2.5\text{ m} - x\operatorname{cosec} 60°)x^2\text{d}x = \frac{2.5^4}{12}\sin^3 60°\text{ m}^4$$

$$\text{First moment} = \text{Area} \times \frac{\text{Depth}}{3}$$

$$= \frac{1}{2}2.5 \times 2.5\sin 60° \times \frac{2.5\sin 60°}{3}\text{ m}^3 = \frac{2.5^3}{6}\sin^2 60°\text{ m}^3$$

$$\therefore \text{ Depth of C.P.} = \frac{2.5}{2}\sin 60°\text{ m} = \frac{\text{Depth}}{2}$$

∴ Thrust is equally divided between XX and bottom.

Thrust = Area × Pressure at centroid

$$= \frac{1}{2}2.5^2\sin 60° \times 1000 \times 9.81 \times \frac{2.5\sin 60°}{3}\text{N} = 19\,160\text{ N}$$

∴ Load at bottom = **9580 N**; at each upper corner **4790 N**

2.12 Let shaft be at depth h below free surface. Then force on disc

$= \pi R^2 \varrho g h$.

By parallel axes theorem, 2nd moment of area about free surface $= \pi R^4/4 + \pi R^2 h^2$.

1st moment of area about free surface $= \pi R^2 h$

∴ Depth of C.P. $= \dfrac{R^2}{4h} + h$ below free surface, that is, $R^2/4h$ below shaft

∴ Turning moment on shaft

$$= \pi R^2 \varrho g h \times \frac{R^2}{4h} = \frac{\pi R^4 \varrho g}{4} \quad \text{[independent of } h\text{]}$$

$$= \frac{\pi(0.6\text{ m})^4 1000\text{ kg}\cdot\text{m}^{-3} \times 9.81\text{ N}\cdot\text{kg}^{-1}}{4} = \mathbf{999\ N\cdot m}$$

2.13

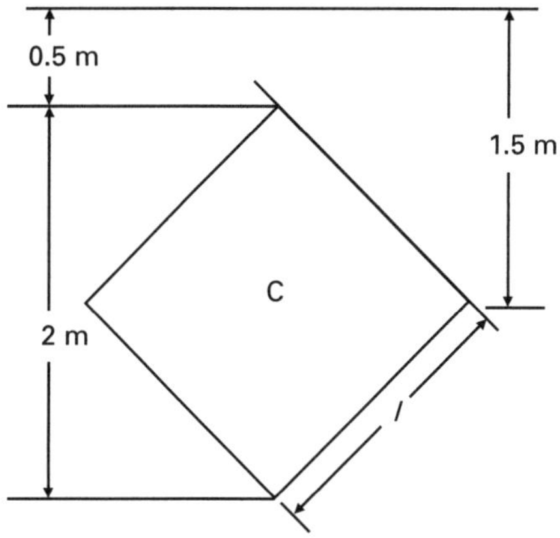

Force on plate

$= 1150\ \text{kg}\cdot\text{m}^{-3}$

$\times 9.81\ \text{N}\cdot\text{kg}^{-1}$

$\times 1.5\ \text{m}(\sqrt{2}\ \text{m})^2$

$= \mathbf{33.84\ kN}$

$$(Ak^2)_{c,\,\perp\,\text{plate}} = \frac{Al^2}{6}$$

$$\therefore\ (Ak^2)_{c,\,\text{diagonal}} = \frac{Al^2}{12}$$

since diagonals are perpendicular

$\therefore$ Depth of C.P. below free surface

$$= \frac{(Al^2/12) + A\bar{y}^2}{A\bar{y}} = \bar{y} + \frac{l^2}{12\bar{y}} = \left\{1.5 + \frac{(\sqrt{2})^2}{12 \times 1.5}\right\} \text{m}$$

$= 1.611$ m, that is, **1.111 m from top of aperture**

$\therefore$ Total moment about hinge $= 33.84\ \text{kN} \times 1.111/\sqrt{2}\ \text{m}$

$= \mathbf{26.59\ kN\cdot m}$

2.14 Width of gates = (3 m) sec 30° = 3.464 m

Thrust on 'deep' side of gate

$= (1000 \times 9.81 \times 4.5)(9 \times 3.464)\text{N} = 1.376\ \text{MN}$

Trust on 'shallow' side of gate

$= (1000 \times 9.81 \times 1.35)(2.7 \times 3.464)\text{N}$

$= \underline{0.124\ \text{MN}}$

Net thrust = (1.376 – 0.124) MN = 1.252 MN

$$\therefore\ \text{Force between gates} = \frac{1.252\ \text{MN}}{2\sin 30^\circ} = \mathbf{1.252\ MN}$$

Resultant force F acts at height y given by

$F_1\dfrac{h_1}{3} - F_2\dfrac{h_2}{3} = Fy$, since F_1, F_2 act at $\dfrac{2}{3}h_1, \dfrac{2}{3}h_2$ below free surfaces

$$\therefore\ y = \frac{1.376 \times 9/3 - 0.124 \times 2.7/3}{1.252}\ \text{m} = 3.208\ \text{m}$$

Total hinge reaction R also acts at this height.

If top hinge is distance x above bottom hinge,

$R_{top}x = R(3.208 - 0.6)$ m

$\therefore\ x = \dfrac{R}{R_{top}} 2.608 \text{ m} = 3 \times 2.608 \text{ m} = 7.82 \text{ m},$

that is, **8.42 m above base**

2.15

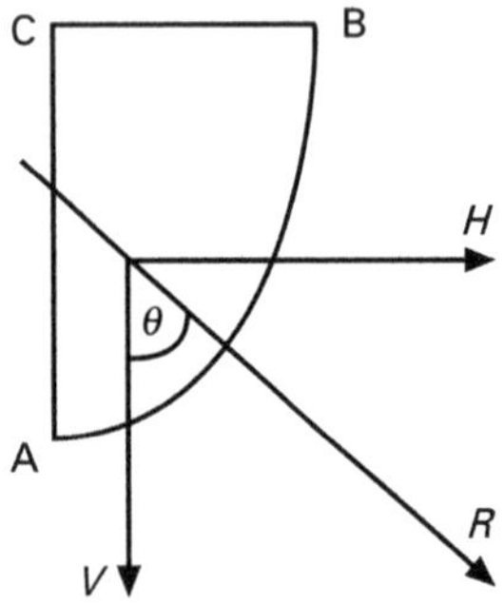

Horizontal component H

$= $ Thrust on vertical projection AC divided by width

$= \dfrac{1}{2} 1000 \times 9.81 \times 27^2 \text{ N}\cdot\text{m}^{-1}$

$= 3.576 \text{ MN}\cdot\text{m}^{-1}$ acting at $\dfrac{2}{3} \times 27$ m

$= 18$ m below BC

Vertical component $V =$ Weight of water ABC

$$\text{Area ABC} = \int_0^{27} x\text{d}y = \sqrt{18} \int_0^{27} y^{1/2}\text{d}y = \frac{2}{3}\sqrt{18}(27)^{3/2} \text{ m}^2$$

$$= 396.8 \text{ m}^2$$

$\therefore$ Vertical component $= 1000 \times 9.81 \times 396.8 \text{ N}\cdot\text{m}^{-1}$

$= 3.893 \text{ MN}\cdot\text{m}^{-1}$

It acts through centroid of ABC. Moments of area about AC:

$$396.8\bar{x} = \int_0^{27} x\text{d}y\frac{x}{2} = \int_0^{27} 9y\text{d}y = 9 \times \frac{27^2}{2} \text{ m}^3$$

whence $\bar{x} = 8.27$ m

$\theta = \arctan \dfrac{3.576}{3.893} = \mathbf{42.57^\circ}$

Resultant $= \sqrt{(3.576^2 + 3.893^2)} \text{ MN}\cdot\text{m}^{-1} = \mathbf{5.29\ MN\cdot m^{-1}}$

It intersects free surface at $(18\tan\theta - 8.27)$ m from C

$= 8.27$ m from C

that is $\{8.27 + \sqrt{(18 \times 27)}\}$ m

$= \mathbf{30.31\ m\ from\ B}$

2.16

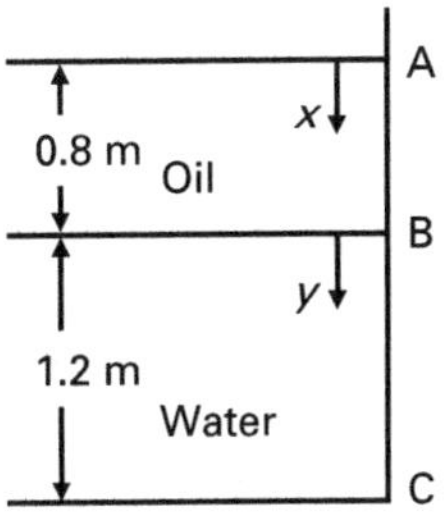

Relevant forces are only those on vertical plane 0.5 m wide.

Total force on AB $= F_1$

$$= \int_0^{0.8\ \text{m}} \varrho_{\text{oil}} g x b \,\text{d}x = \left[\frac{1}{2}\varrho_{\text{oil}} g b x^2\right]_0^{0.8\text{m}}$$

$$= \frac{1}{2} 850\ \text{kg}\cdot\text{m}^{-3} \times 9.81\ \text{N}\cdot\text{kg}^{-1} \times 0.5\ \text{m}(0.8\ \text{m})^2 = 1334\ \text{N}$$

$$\text{Total force on BC} = \int_0^{1.2\ \text{m}} (\varrho_{\text{oil}} g 0.8\ \text{m} + \varrho_{\text{water}} g y) b\ \text{d}y$$

$$= b\varrho_{\text{water}} g \int_0^{1.2\ \text{m}} (0.85 \times 0.8\ \text{m} + y)\text{d}y$$

$$= b\varrho_{\text{water}} g \left[(0.68\ \text{m})y + \frac{y^2}{2}\right]_0^{1.2\text{m}}$$

$$= 0.5 \times 1000 \times 9.81(0.816 + 0.72)\ \text{N}$$

$$= 7535\ \text{N}$$

Total force $F = (7535 + 1334)\text{N} = \mathbf{8869\ N}$

Let total force act at height z above base of tank. Then moments about axis through C:

$$Fz = F_1\left(1.2 + \frac{0.8}{3}\right)\text{m} + \int_0^{1.2\text{m}} b\varrho_{\text{water}} g(0.68\ \text{m} + y)(1.2\ \text{m} - y)\text{d}y$$

$$= 1334 \times 1.467\ \text{N}\cdot\text{m} + 0.5 \times 1000 \times 9.81\ \text{N}\cdot\text{m}^{-2} \times \left[(0.816\ \text{m}^2)y + \frac{0.52\ \text{m}}{2}y^2 - \frac{y^3}{3}\right]_0^{1.2\text{m}}$$

$$= 1957\ \text{N}\cdot\text{m} + 500 \times 9.81[0.9792 + 0.3744 - 0.576]\ \text{N}\cdot\text{m}$$

$$= 5771\ \text{N}\cdot\text{m}$$

$$\therefore\ z = \frac{5771}{8869}\ \text{m} = \mathbf{0.651\ m}$$

2.17

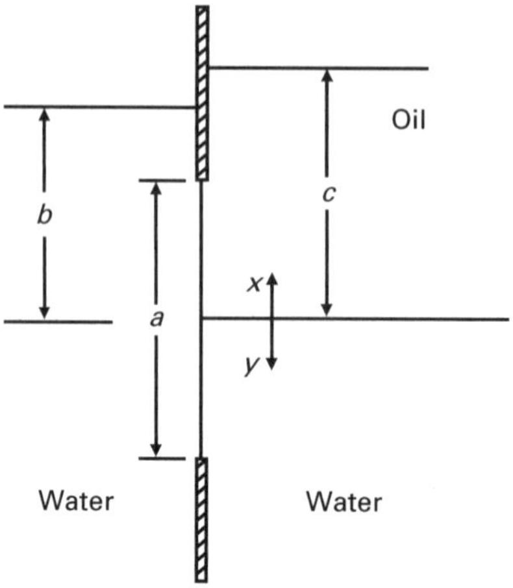

$\overrightarrow{\text{Force}} = \varrho g b a^2$

$\overleftarrow{\text{Force}} = \sigma \varrho g \left(c - \frac{a}{4}\right) \frac{a^2}{2}$

$+ \left(\sigma \varrho g c + \varrho g \frac{a}{4}\right) \frac{a^2}{2}$

$= \sigma \varrho g c a^2 + \frac{\varrho g a^3}{8}$

$\times (1 - \sigma)$

For zero net force

$\boldsymbol{b = \sigma c + \frac{a}{8}(1 - \sigma)}$

Total moment ↶ about centre-line for forces on left

$$= -\int_0^{a/2} \varrho g(b - x)ax \, dx + \int_0^{a/2} \varrho g(b + y)ay \, dy$$

$$= \varrho g a \left(-\frac{ba^2}{8} + \frac{a^3}{24} + \frac{ba^2}{8} + \frac{a^3}{24}\right) = \frac{1}{12}\varrho g a^4$$

$\therefore$ Net $\overrightarrow{\text{Force}}$ acts at $\frac{1}{12}\varrho g a^4 \div \varrho g b a^2 = a^2/12b$ below centre-line.

Total moment ↷ about centre-line for forces on right

$$= -\int_0^{a/2} \sigma \varrho g(c - x)ax \, dx + \int_0^{a/2} (\sigma \varrho g c + \varrho g y)ay \, dy$$

$$= \varrho g a \left(-\sigma c \frac{a^2}{8} + \sigma \frac{a^3}{24} + \sigma c \frac{a^2}{8} + \frac{a^3}{24}\right) = \frac{1}{24}\varrho g a^4(1 + \sigma)$$

$\therefore$ Net $\overleftarrow{\text{force}}$ acts at $\frac{1}{24}\varrho g a^4(1 + \sigma) \div \varrho g b a^2$

$= a^2(1 + \sigma)/24b$ below centre-line.

$\therefore$ Axis of couple is $\frac{1}{2}\left\{\frac{a^2}{12b} + \frac{a^2(1 + \sigma)}{24b}\right\}$

$= \boldsymbol{\frac{a^2}{48b}(3 + \sigma)}$ **below centre-line**

2.18 Pressure at centroid $= (15\,000 + 900 \times 9.81 \times 1)$ Pa $= 23\,829$ Pa

$\therefore$ Total force $= 23\,829$ Pa $\times 0.24$ m$^2 = 5719$ N

This acts on vertical centre-line

$\therefore$ Force on lock $= \dfrac{1}{2} \times 5719\ \text{N} = \mathbf{2859\ N}$

Air pressure is equivalent to $\dfrac{15\,000}{900 \times 9.81}\ \text{m} = 1.699$ m of oil

$\therefore$ Equivalent free (atmospheric) surface is at 2.699 m above centre-line

$\therefore$ Depth of C.P. below centre-line $= (Ak^2)_c/A\bar{y}$

$$= \frac{(0.4\ \text{m})^2}{12} \Big/ 2.699\ \text{m} = 0.00494\ \text{m}$$

Moments about horizontal axis through upper hinge:

$5719(0.125 + 0.00494)\ \text{N}\cdot\text{m} = 2859 \times 0.125\ \text{N}\cdot\text{m} + F_L(0.25\ \text{m})$

$\therefore$ Force on lower hinge $= F_L = \mathbf{1543\ N}$

and force on upper hinge $= (2859 - 1543)\ \text{N} = \mathbf{1317\ N}$

2.19 2.7 kg of iron occupy $\dfrac{2.7\ \text{kg}}{7500\ \text{kg}\cdot\text{m}^{-3}} = 0.00036\ \text{m}^3$

$\therefore$ Buoyancy force $= 0.00036\ \text{m}^3 \times 1000\ \text{kg}\cdot\text{m}^{-3} \times 9.81\ \text{N}\cdot\text{kg}^{-1}$

$= 0.36 \times 9.81$ N

$\therefore$ Spring balance reads $(2.7 - 0.36)$ kgf $= \mathbf{2.34\ kgf}$

Parcel balance reads $(5 + 0.36)$ kgf $= \mathbf{5.36\ kgf}$

2.20

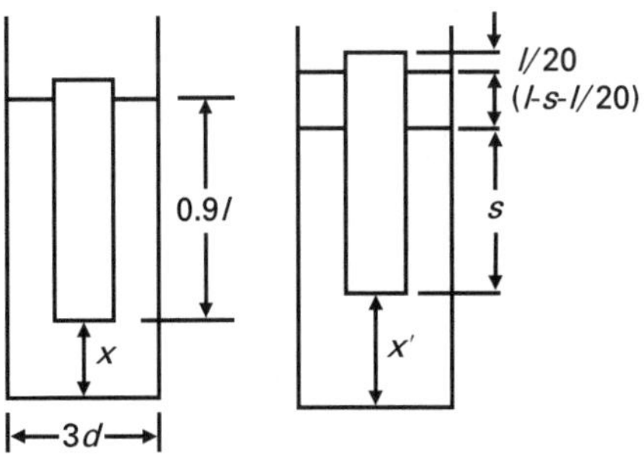

Archimedes for case II:

$$0.9l = 1 \times s + 0.8\left(\frac{19}{20}l - s\right)$$

whence $s = 0.7l$

Volume of water is constant

$$\therefore \frac{\pi}{4}(3d)^2 x + 0.9l \times \left\{\frac{\pi}{4}(3d)^2 - \frac{\pi}{4}d^2\right\} = \frac{\pi}{4}(3d)^2 x' + 0.7l\left\{\frac{\pi}{4}(3d)^2 - \frac{\pi}{4}d^2\right\}$$

$\therefore\ 9x + 0.9l\{9 - 1\} = 9x' + 0.7l\{9 - 1\} \quad \therefore\ \mathbf{x' - x = 0.1778}\boldsymbol{l}$

2.21

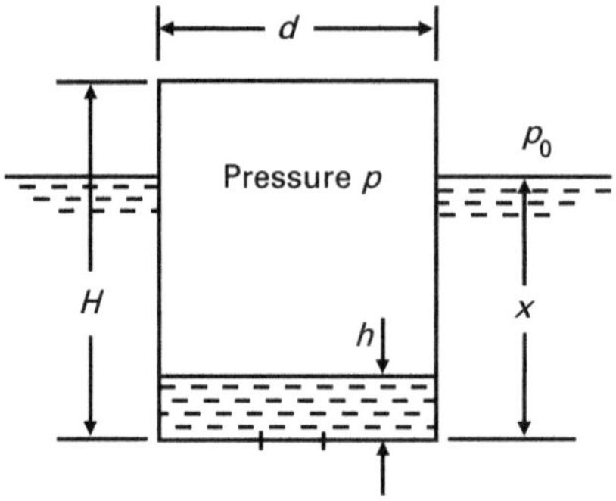

Archimedes $\frac{\pi}{4}d^2(x-h)\varrho g$

$= 27 \times 9.81$ N

At base of cylinder, pressure

$= p_0 + \varrho g x = p + \varrho g h$

$\therefore p - p_0 = \varrho g(x-h)$

$$= \frac{27 \times 9.81 \text{ N}}{(\pi/4)(0.3 \text{ m})^2}$$

$= \mathbf{3747\ Pa}$

For isothermal compression pV = constant

$\therefore\ p(H-h) = p_0 H$

$$\therefore\ h = \frac{p-p_0}{p}H = \frac{3747}{105\,047} \times 450 \text{ mm} = \mathbf{16.05\ mm}$$

$$x - h = \frac{27 \times 9.81 \text{ N}}{(\pi/4)(0.3 \text{ m})^2 1000 \text{ kg}\cdot\text{m}^{-3} \times 9.81 \text{ N}\cdot\text{kg}^{-1}} = 0.382 \text{ m}$$

$\therefore \boldsymbol{x} = \mathbf{398\ mm}$

2.22 From eqn 2.7, p at 6000 m is $p_0\left(1 - \frac{\lambda z}{T_0}\right)^{g/R\lambda}$

$$= 101 \text{ kPa}\left(1 - \frac{0.0065 \times 6000}{288.15}\right)^{9.81/287\times0.0065} = 47.01 \text{ kPa}$$

$$\therefore\ \varrho \text{ at 6000 m is } \frac{47\,010}{287(288.15 - 0.0065 \times 6000)} \text{ kg}\cdot\text{m}^{-3}$$

$= 0.6574 \text{ kg}\cdot\text{m}^{-3}$

which must be same as effective density of balloon.

$\therefore$ Total mass of balloon $= 0.6574 \times \frac{\pi}{6}0.8^3$ kg $= 0.17625$ kg

$\therefore$ Mass of helium = (176.25 − 160)g = **16.25 g**

2.23 $\text{BM} = Ak^2/V = \frac{\pi}{64}d^4 \Big/ \left(\frac{\pi}{4}d^2 \times 0.6l\right) = d^2/9.6l$

B is at $0.3l$ above base.

$\therefore$ When M and G coincide, BM $= 0.2l$

$\therefore d^2 = 0.2 \times 9.6l^2 \quad \therefore d/l = \sqrt{1.92} = \mathbf{1.386}$

2.24 Weight of pontoon

$$= (6 \times 3 \times 0.9)\ \mathrm{m}^3 \times 1000\ \mathrm{kg \cdot m^{-3}} \times 9.81\ \mathrm{N \cdot kg^{-1}} = 158.9\ \mathrm{kN}$$

$$\mathrm{BM} = Ak^2/V = [(6 \times 3^3/12)/(6 \times 3 \times 0.9)]\ \mathrm{m} = 0.833\ \mathrm{m}$$

$$\mathrm{GM} = \left(0.833 + \frac{0.9}{2} - 0.7\right)\ \mathrm{m} = 0.583\ \mathrm{m}$$

$$7600\ \mathrm{N \cdot m} = W(\mathrm{GM}) \sin\theta$$

$$\therefore \sin\theta = \frac{7600}{158.9 \times 10^3 \times 0.583} \qquad \therefore \theta = \mathbf{4.70^\circ}$$

2.25 If relative density $= \sigma$, depth of immersion $h = 150\sigma$ mm
$\therefore$ Height of B $= 75\sigma$ mm

$$\mathrm{BM} = Ak^2/V = \frac{\pi}{64}d^4 \Big/ \frac{\pi}{4}d^2h = \frac{d^2}{16h} = \frac{75^2}{16 \times 150\sigma}\ \mathrm{mm}$$

$$= \frac{75}{32\sigma}\ \mathrm{mm}$$

For stability $\quad \mathrm{BM} > \mathrm{BG} \quad \therefore \dfrac{75}{32\sigma} > \dfrac{150}{2} - 75\sigma$

that is $\dfrac{1}{32\sigma} > 1 - \sigma$

$$\therefore 32\sigma^2 - 32\sigma + 1 > 0 \qquad \therefore \sigma > \frac{16 + \sqrt{256 - 32}}{32} = 0.9677$$

$$\text{or } \sigma < \frac{16 - \sqrt{256 - 32}}{32} = 0.0323$$

this is unrealistic since cylinder is solid

$\therefore 0.9677 < \sigma < 1.0$

Mass of equal volume of water

$$= \frac{\pi}{4}(0.075)^2 \times 0.15\ \mathrm{m}^3 \times 1000\ \mathrm{kg \cdot m^{-3}} = 0.663\ \mathrm{kg}$$

$\therefore$ Mass of cylinder is **between 0.641 kg and 0.663 kg**

2.26 $\mathrm{Torque} = \dfrac{\mathrm{Power}}{\omega} = \dfrac{3.34 \times 10^6}{1.4 \times 2\pi}\ \mathrm{N \cdot m} = W(\mathrm{GM}) \sin\theta$

$$= 80 \times 10^6 (\mathrm{G_1M_1}) \sin 0.53^\circ$$

whence $\mathrm{G_1M_1} = 0.513$ m

$\therefore \mathrm{B_1M_1} = (0.513 + 1.6 - 0.3)\ \mathrm{m} = 1.813\ \mathrm{m}$

$$B_1M_1 \times V_1 = Ak^2 = B_2M_2 \times V_2$$

$$\therefore B_2M_2 = 1.813 \text{ m} \times \frac{80 \times 10^6}{80 \times 10^6 - 400 \times 10^3 \times 9.81} = 1.907 \text{ m}$$

$$\frac{3.34 \times 10^6}{1.4 \times 2\pi} = 76.076 \times 10^6 (G_2M_2) \sin 0.75^\circ$$

whence $G_2M_2 = 0.3813$ m

$$\therefore B_2G_2 = (1.907 - 0.381) \text{ m} = 1.525 \text{ m}$$

$$B_2G_1 = (1.6 - 0.3 + 0.075) \text{ m} = 1.375 \text{ m}$$

$$\therefore G_1G_2 = (1.525 - 1.375) \text{ m} = \mathbf{0.150 \text{ m}}$$

2.27 Volume of water displaced $= \dfrac{355}{1025}\text{m}^3 = 0.3463 \text{ m}^3 = \dfrac{1}{3}\pi r^2 h$

$$= \frac{1}{3}\pi r^3 \sqrt{3}$$

$\therefore r^3 = 0.1910 \text{ m}^3 \quad \therefore r = 0.576$ m

$$BM = Ak^2/V = \frac{\pi}{4}r^4 \Big/ \frac{1}{3}\pi r^2 h = \frac{3}{4}\frac{r^2}{h} = \frac{r}{4}\sqrt{3} = 0.2493 \text{ m}$$

B is at $\dfrac{3}{4}h = 0.748$ m above vertex, that is $(0.6\sqrt{3} - 0.748)$ m

$= 0.2912$ m below top

$\therefore$ M is $(0.2912 - 0.2493)$ m $= 0.0419$ m below top – this is limiting position of G.

Let beacon be x metres above top. Then moments about axis in top:

$$300(0.6\sqrt{3} - 0.75) - 55x = 355 \times 0.0419,$$

whence $\mathbf{x = 1.308 \text{ m}}$

2.28 Depth of immersion $= 0.85 \times 0.8$ m $= 0.68$ m

$\therefore$ B is 0.34 m above base

$$BM = Ak^2/V = \frac{\pi}{64}(1 \text{ m})^4 \Big/ \frac{\pi}{4}(1 \text{ m})^2 \times 0.68 \text{ m} = 0.0919 \text{ m}$$

$$\therefore GM = (0.0919 + 0.34 - 0.4) \text{ m} = 0.0319 \text{ m}$$

$$t = 2\pi\sqrt{\frac{k^2}{g(GM)}} = 2\pi\sqrt{\frac{l^2/12 + r^2/4}{g(GM)}} = 2\pi\sqrt{\frac{0.8^2/12 + 0.5^2/4}{9.81 \times 0.0319}} \text{ s}$$

$$= \mathbf{3.822 \text{ s}}$$

2.29 $0.405\ \text{m}^3 = \left\{0.9^3 - 0.9\frac{0.9}{2}0.9\tan(\phi+\theta)\right\}\text{m}^3$

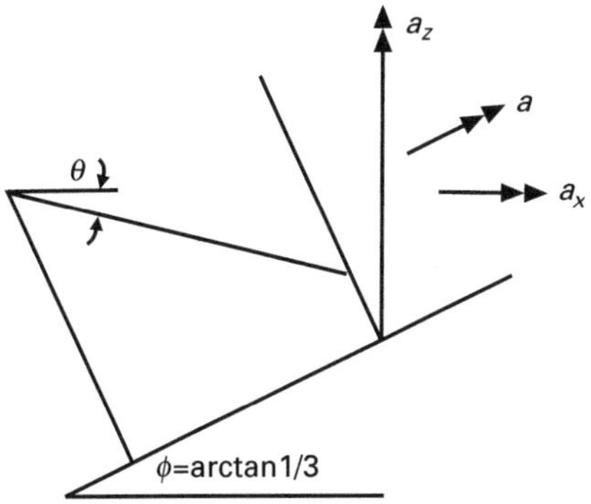

$$\text{whence } \tan(\phi+\theta) = \frac{8}{9}$$

$$\therefore \tan\theta = \frac{8/9 - \tan\phi}{1 + \frac{8}{9}\tan\phi} = \frac{3}{7}$$

$$\tan\theta = \frac{3}{7} = \frac{a_x}{a_z + g}$$

$$= \frac{a\cos\phi}{a\sin\phi + g}$$

$$\text{whence } a = g\sqrt{10}/6$$

Total mass $= (340 + 0.405 \times 850)\ \text{kg} = 684.25\ \text{kg}$

$\therefore F = 684.25 \times 9.81\sqrt{10}/6\ \text{N} = \mathbf{3538\ N}$

2.30

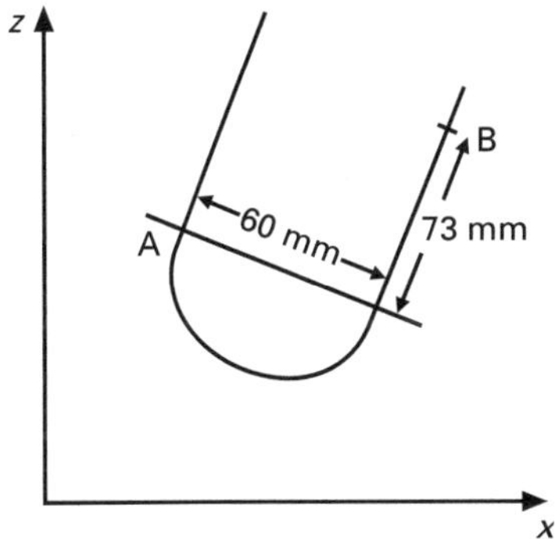

$a_x = 2\cos 20°\ \text{m}\cdot\text{s}^{-2}$;
$a_z = -2\sin 20°\ \text{m}\cdot\text{s}^{-2}$
If A is origin, B is at
$\{(60\cos 20° + 73\sin 20°)\ \text{mm},$
$(73\cos 20° - 60\sin 20°)\ \text{mm}\}$
that is (81.35 mm, 48.08 mm)

Pressure at B $= -\varrho a_x x - \varrho(g + a_z)z + \text{constant}$
If p at A is taken as zero, constant = 0

$$\therefore p_B = -790\ \text{kg}\cdot\text{m}^{-3} \times \{2\cos 20° \times 0.08135 + (9.81 - 2\sin 20°)0.04808\}\ \text{m}^2\cdot\text{s}^{-2}$$
$$= -790(0.1529 + 0.4388)\ \text{Pa} = \mathbf{-467\ Pa}$$

Chapter 3

3.1 $$u_B = \frac{0.3}{0.15} \times 1.8\ \text{m}\cdot\text{s}^{-1} = 3.6\ \text{m}\cdot\text{s}^{-1}$$

$$p_B = p_A + \frac{1}{2}\varrho u_A^2 + \varrho g z_A - \varrho g z_B - \frac{1}{2}\varrho u_B^2$$

$$= \Big\{1.17 \times 10^5 + \frac{1000}{2}(1.8^2 - 3.6^2) - 1000 \times 9.81 \times 6\Big\}\ \text{Pa gauge} = \mathbf{53.3\ kPa\ gauge}$$

3.2

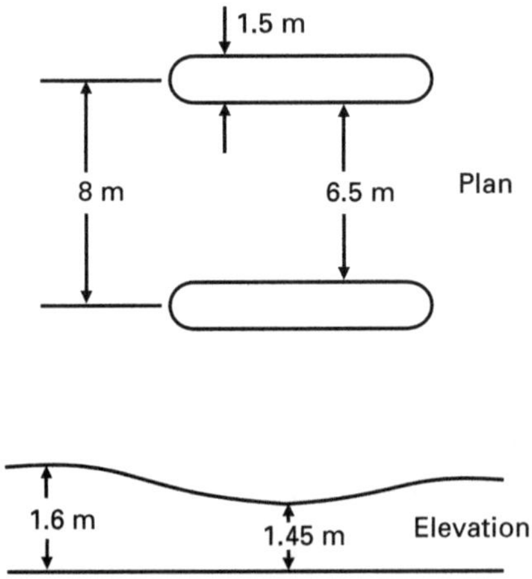

For any given streamline

$$\frac{u_1^2}{2g} + \left(\frac{p_1}{\varrho g} + z_1\right) = \frac{u_2^2}{2g} + \left(\frac{p_2}{\varrho g} + z_2\right)$$

If pressure variation with depth is same as hydrostatic – true only if there is no appreciable acceleration ⊥ streamlines, that is, streamlines must be sensibly straight and parallel where measurements are made – then $(p/\varrho g) + z$ = depth of stream at this point.

$$\therefore\ \frac{1}{2g}\left(\frac{Q}{8 \times 1.6\ \text{m}^2}\right)^2 + 1.6\ \text{m} = \frac{1}{2g}\left(\frac{Q}{6.5 \times 1.45\ \text{m}^2}\right)^2 + 1.45\ \text{m}$$

whence $\mathbf{Q = 23.9\ m^3\cdot s^{-1}}$

3.3 $$q = \frac{p_2 - p_1}{\varrho} + \frac{1}{2}\left(u_2^2 - u_1^2\right) + g\,(z_2 - z_1) + c\,(T_2 - T_1)$$

$$= \frac{(5.5 \times 10^6 + 1.225 \times 9.81 \times 600)\ \text{Pa}}{10^3\ \text{kg}\cdot\text{m}^{-3}} + \frac{1}{2}(2^2 - 0)\ \text{m}^2\cdot\text{s}^{-2}$$

$$+ 9.81\ \text{m}\cdot\text{s}^{-2}(0 - 600)\ \text{m} + 4.187 \times 10^3\ \text{J}\cdot\text{kg}^{-1}\cdot\text{K}^{-1}(1.8\ \text{K})$$

$$= 7159.81\ \text{m}^2\cdot\text{s}^{-2}$$

$$\varrho Q = 1000\ \text{kg}\cdot\text{m}^{-3} \times \frac{\pi}{4}1.2^2\ \text{m}^2 \times 2\ \text{m}\cdot\text{s}^{-1} = 2262\ \text{kg}\cdot\text{s}^{-1}$$

$\therefore$ Heat flow $= 2262 \times 7159.81$ W $=$ **16.20 MW**

3.4 $u = u_0\left(1 + \frac{y}{h}\right)$ where $y =$ height above base, $h =$ full depth,

$u_0 =$ velocity at base

$$\frac{\text{K.E. flow}}{\text{width}} = \int_0^h \varrho u\ \text{d}y \frac{u^2}{2} = \int_0^h \frac{\varrho}{2}u_0^3\left(1 + \frac{y}{h}\right)^3 \text{d}y$$

$$= \frac{1}{8}\varrho u_0^3 h\left[\left(1 + \frac{y}{h}\right)^4\right]_0^h = \frac{15}{8}\varrho u_0^3 h$$

$$\therefore \text{K.E. per unit mass} = \frac{15}{8}\varrho u_0^3 h \div \varrho\frac{3}{2}u_0 h = \frac{5}{2}\frac{u_0^2}{2}$$

$$\text{since mean velocity} = \frac{3}{2}u_0$$

$$\therefore \alpha = \frac{5}{2}\frac{u_0^2}{2} \div \frac{\overline{u}^2}{2} = \frac{5}{2} \div \left(\frac{3}{2}\right)^2 = \mathbf{\frac{10}{9}}$$

3.5

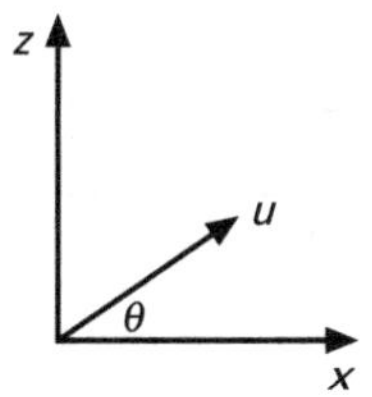

$$x = (u\cos\theta)t; \quad z = (u\sin\theta)t - \frac{1}{2}gt^2$$

$$\therefore\ z = x\tan\theta - \frac{gx^2}{2u^2\cos^2\theta}$$

$$\text{whence} \quad u^2 = \frac{gx^2}{2\cos^2\theta(x\tan\theta - z)}$$

Minimum u^2 requires max $\cos^2\theta(x\tan\theta - z) = f$, say.

$$\frac{\text{d}f}{\text{d}\theta} = -2\cos\theta\sin\theta(x\tan\theta - z) + \cos^2\theta(x\sec^2\theta)$$

$$= (1 - 2\sin^2\theta)x + z\sin 2\theta$$

$$= x\cos 2\theta + z\sin 2\theta = 0 \quad \text{when}\ \tan 2\theta = -x/z$$

$$\frac{\text{d}^2 f}{\text{d}\theta^2} = -2x\sin 2\theta + 2z\cos 2\theta = 2\cos 2\theta(z - x\tan 2\theta)$$

When $\tan 2\theta = -x/z$, $\dfrac{d^2 f}{d\theta^2} = 2\cos 2\theta\left(z + \dfrac{x^2}{z}\right)$

which is negative since $\cos 2\theta$ is negative

$\therefore f$ is then a maximum.

$$\tan 2\theta = -\frac{20}{15} = -\frac{4}{3} \quad \therefore \cos^2\theta = \frac{1+\cos 2\theta}{2} = \frac{1-3/5}{2} = \frac{1}{5}$$

$\therefore \tan\theta = 2$

$\therefore \theta =$ **63.43° to horizontal**

$$u^2 = \frac{gx^2}{2\cos^2\theta(x\tan\theta - z)} = \frac{9.81 \times 20^2}{2 \times \frac{1}{5}(20 \times 2 - 15)}\ \mathrm{m^2 \cdot s^{-2}}$$

$$= 392.4\ \mathrm{m^2 \cdot s^{-2}}$$

$$p = \varrho g h = \varrho g \frac{u^2}{2gC_v^2} = \frac{\varrho u^2}{2C_v^2} = \frac{1000 \times 392.4}{2 \times 0.95^2}\ \mathrm{kg \cdot m^{-1} \cdot s^{-2}}$$

$$= \mathbf{217.4 \times 10^3\ Pa}$$

3.6 Δp across orifice $= 0.271\ \mathrm{m} \times 0.1 \times (800 \times 9.81)\ \mathrm{N \cdot m^{-3}}$
$= 212.7\ \mathrm{Pa}$

$$\varrho = \frac{p}{RT} = \frac{0.772\ \mathrm{m}(13\,560 \times 9.81)\ \mathrm{N \cdot m^{-2}}}{287\ \mathrm{J \cdot kg^{-1} \cdot K^{-1}} \times 288.95\ \mathrm{K}} = 1.238\ \mathrm{kg \cdot m^{-3}}$$

$$\therefore\ Q = C_d \frac{\pi}{4} d^2 \sqrt{2\frac{\Delta p}{\varrho}} = 0.602 \frac{\pi}{4}(0.05\ \mathrm{m})^2 \sqrt{\frac{2 \times 212.7\ \mathrm{Pa}}{1.238\ \mathrm{kg \cdot m^{-3}}}}$$

$$= \mathbf{0.02191\ m^3 \cdot s^{-1}}$$

3.7 $$C_c = \left(\frac{39.5}{50}\right)^2 = 0.6241 \quad C_d = \frac{0.018\ \mathrm{m^3 \cdot s^{-1}}}{\frac{\pi}{4}(0.05\ \mathrm{m})^2 \sqrt{\dfrac{2 \times 10^5\ \mathrm{Pa}}{850\ \mathrm{kg \cdot m^{-3}}}}} = 0.598$$

$\therefore\ C_v = C_d / C_c =$ **0.958**

3.8 Static pressure $= 1026\ \mathrm{kg \cdot m^{-3}} \times 9.81\ \mathrm{N \cdot kg^{-1}} \times 15\ \mathrm{m}$
$= 1.510 \times 10^5\ \mathrm{Pa}$

$$\frac{1}{2}\varrho u^2 = \frac{1}{2} 1026\ \mathrm{kg \cdot m^{-3}} \left(\frac{16 \times 10^3}{3600}\ \mathrm{m \cdot s^{-1}}\right)^2$$

$$= 1.013 \times 10^4\ \mathrm{Pa}$$

$\therefore$ Stagnation pressure $= 1.611 \times 10^5\ \mathrm{Pa} =$ **161.1 kPa gauge**

3.9 Theoretical $u_1 = \dfrac{(0.040/0.96)\ \mathrm{m^3\cdot s^{-1}}}{(\pi/4)(0.15\ \mathrm{m})^2} = 2.358\ \mathrm{m\cdot s^{-1}}$

and $u_2 = 4u_1$

$$
\begin{aligned}
p_1 - p_2 &= \frac{1}{2}\varrho\left(u_2^2 - u_1^2\right) + \varrho g(z_2 - z_1) \\
&= 400\ \mathrm{kg\cdot m^{-3}}(15 \times 2.358^2)\ \mathrm{m^2\cdot s^{-2}} \\
&\quad + 800\ \mathrm{kg\cdot m^{-3}} \times 9.81\ \mathrm{N\cdot kg^{-1}} \times 0.15\ \mathrm{m} \\
&= \mathbf{34\,530\ Pa}
\end{aligned}
$$

Manometer measures difference of *piezometric* pressure

$$
\begin{aligned}
&= \frac{1}{2}\varrho\left(u_2^2 - u_1^2\right) = 33\,360\ \mathrm{Pa} \\
&= (\varrho_{\mathrm{Hg}} - \varrho_{\mathrm{liq}})gh
\end{aligned}
$$

$$\therefore\ h = \frac{33\,360\ \mathrm{Pa}}{(13\,560 - 800)9.81\ \mathrm{N\cdot m^{-3}}} = \mathbf{0.2665\ m}$$

3.10 Net effective area of piston $= \dfrac{\pi}{4}(0.1^2 - 0.02^2)\ \mathrm{m^2} = 0.00754\ \mathrm{m^2}$

$\therefore$ Pressure difference required $= 180\ \mathrm{N}/0.00754\ \mathrm{m^2}$

$= 23\,870\ \mathrm{Pa}$

$$u_1 = \frac{0.15\ \mathrm{m^3\cdot s^{-1}}}{(\pi/4)(0.35\ \mathrm{m})^2} = 1.559\ \mathrm{m\cdot s^{-1}} \qquad \therefore\ u_2 = 1.559n\ \mathrm{m\cdot s^{-1}}$$

where n = area ratio

$$p_1 - p_2 = \frac{1}{2}\varrho\left(u_2^2 - u_1^2\right)$$

$$\therefore\ 23\,870\ \mathrm{Pa} = \frac{1}{2}950\ \mathrm{kg\cdot m^{-3}} \times 1.559^2(n^2 - 1)\ \mathrm{m^2\cdot s^{-2}}$$

whence $n = 4.66$ and throat diameter $= \dfrac{350\ \mathrm{mm}}{\sqrt{4.66}}$

$= \mathbf{162.2\ mm}$

3.11

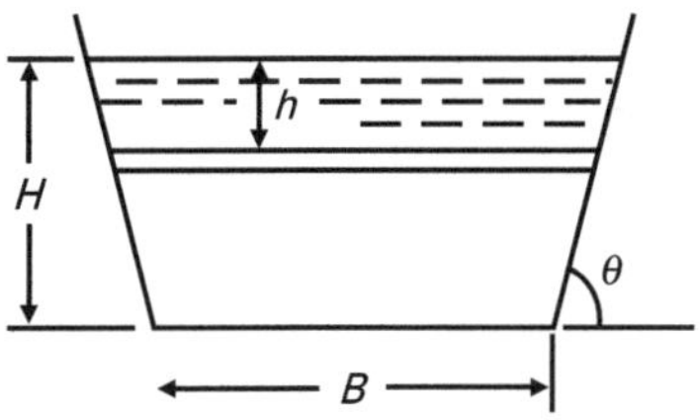

$\tan\theta = \dfrac{300}{200} = 1.5$

Ideal discharge through strip

$= \{B + 2(H - h)\cot\theta\}$

$\times\ \mathrm{d}h\sqrt{2gh}$

$$\therefore\ Q_{\text{ideal}} = \sqrt{2g}\int_0^H \left\{B + 2(H-h)\frac{2}{3}\right\} h^{1/2}\mathrm{d}h$$
$$= \sqrt{2g}\left[\frac{2}{3}\left(B + \frac{4}{3}H\right)H^{3/2} - \frac{4}{3}\times\frac{2}{5}H^{5/2}\right]$$
$$= \frac{2}{3}\sqrt{2g}H^{3/2}\left(B + \frac{8}{15}H\right)$$

$$\therefore\ Q = \text{this}\ \times C_d = \frac{2}{3}\sqrt{19.62}(0.228)^{3/2}(0.1 + 0.1216)0.6\ \text{m}^3\cdot\text{s}^{-1}$$
$$= \mathbf{0.0427\ m^3\cdot s^{-1}}$$

3.12 $$Q_1 = 0.6\times\frac{2}{3}(1.8\ \text{m})\sqrt{19.62\ \text{m}\cdot\text{s}^{-2}}(0.08\ \text{m})^{3/2} = 0.0722\ \text{m}^3\cdot\text{s}^{-1}$$

$\therefore$ Approach velocity $= 0.0722\ \text{m}^3\cdot\text{s}^{-1} \div 0.3\ \text{m}^2 = 0.2406\ \text{m}\cdot\text{s}^{-1}$

$$\therefore\ \frac{u^2}{2g} = \frac{0.2406^2}{19.62}\ \text{m} = 0.00295\ \text{m}$$

$\therefore$ Better approximation to total head $= (0.08 + 0.00295)$ m
$= 0.08295$ m

$\therefore$ Better approximation to

$$Q = 0.6\times\frac{2}{3}(1.8\ \text{m})\sqrt{19.62\ \text{m}\cdot\text{s}^{-2}}(0.08295\ \text{m})^{3/2}$$
$$= \mathbf{0.0762\ m^3\cdot s^{-1}}$$

Chapter 4

4.1

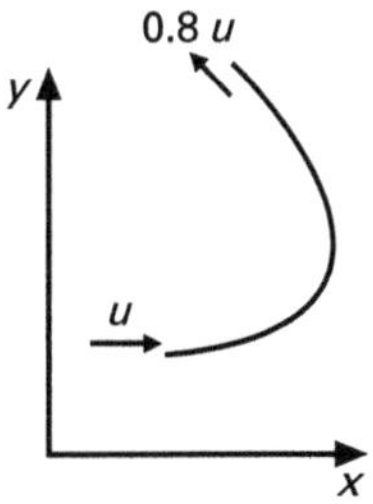

$$F_x = \varrho Q(0.8u\cos 150° - u)$$
$$= \varrho A_1 u^2(-0.4\sqrt{3} - 1)$$
$$F_y = \varrho A_1 u(0.8u\sin 30° - 0) = 0.4\varrho A_1 u^2$$
$$\therefore F^2 = F_x^2 + F_y^2 = \varrho^2 A_1^2 u^4(1.64 + 0.8\sqrt{3})$$

$$\therefore u^2 = \frac{2000\ \mathrm{N}}{(1000\ \mathrm{kg\cdot m^{-3}})(\pi/4)(0.05\ \mathrm{m})^2\sqrt{3.026}} \qquad \therefore u = 24.2\ \mathrm{m\cdot s^{-1}}$$

$$Q = \frac{\pi}{4}(0.05)^2 \times 24.2\ \mathrm{m^3\cdot s^{-1}} = \mathbf{0.0475\ m^3\cdot s^{-1}}$$

4.2

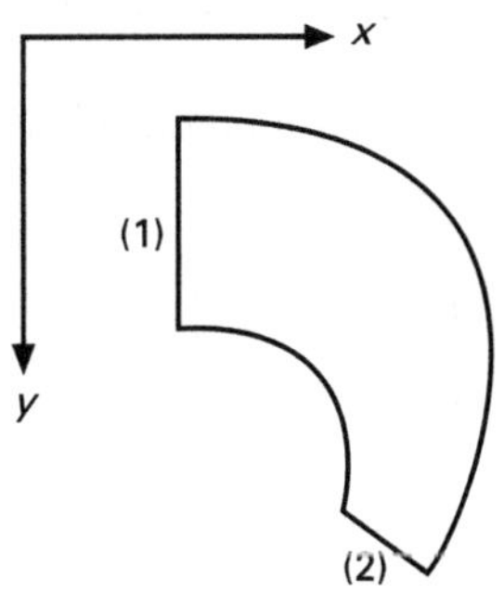

$$u_1 = \frac{0.23}{(\pi/4)(0.3)^2}\ \mathrm{m\cdot s^{-1}} = 3.254\ \mathrm{m\cdot s^{-1}}$$

$$u_2 = 4u_1$$

Energy eqn: $\frac{1.4\times 10^5}{1000\times 9.81}$ m

$$+\frac{3.254^2}{19.62}\ \mathrm{m} + 1.4\ \mathrm{m}$$
$$= \frac{p_2}{1000\times 9.81\ \mathrm{N\cdot m^{-3}}}$$
$$+\frac{16\times 3.254^2}{19.62}\ \mathrm{m}$$

whence $p_2 = 74\,300$ Pa

$$F_x + p_1A_1 + p_2A_2\cos 60° = \varrho Q(u_2\cos 120° - u_1)$$

$$\therefore F_x = 1000\ \text{kg}\cdot\text{m}^{-3} \times 0.23\ \text{m}^3\cdot\text{s}^{-1}$$
$$\times (-4 \times 3.254 \times 0.5 - 3.254)\ \text{m}\cdot\text{s}^{-1}$$
$$- 1.4 \times 10^5\ \text{N}\cdot\text{m}^{-2} \times \frac{\pi}{4}(0.3\ \text{m})^2$$
$$- 74\,300\ \text{N}\cdot\text{m}^{-2}\frac{\pi}{4} \times (0.15\ \text{m})^2 0.5$$
$$= -12\,800\ \text{N}$$

$$F_y - p_2 A_2 \sin 60° + W = \varrho Q(u_2 \sin 120° - 0)$$

$$\therefore F_y = 1000\ \text{kg}\cdot\text{m}^{-3} \times 0.23\ \text{m}^3\cdot\text{s}^{-1} \times 4 \times 3.254 \times \frac{\sqrt{3}}{2}\ \text{m}\cdot\text{s}^{-1}$$
$$+ 74\,300\ \text{N}\cdot\text{m}^{-2} \times \frac{\pi}{4}(0.15\ \text{m})^2\frac{\sqrt{3}}{2}$$
$$- 0.085\ \text{m}^3 \times 1000\ \text{kg}\cdot\text{m}^{-3} \times 9.81\ \text{N}\cdot\text{kg}^{-1}$$
$$= 2896\ \text{N}$$

$F = \sqrt{12\,800^2 + 2896^2}\ \text{N} = \mathbf{13\,120\ N}$

$\tan\theta = \dfrac{2896}{-12\,800} \qquad \therefore \theta = 180° - 12.76°$

Force on *bend* is equal and opposite to this, that is, 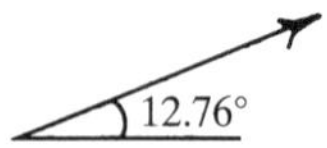

4.3 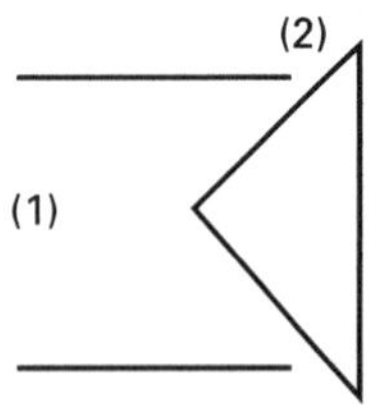

Assume flow at (2) entirely parallel to sides of cone.
Then p_2 = atmospheric. Assuming steady flow and constant ϱ

$$p_1 + \frac{1}{2}\varrho u_1^2 = p_2 + \frac{1}{2}\varrho u_2^2$$

$$\therefore p_1 = \text{atmos.} + \frac{1}{2}1.22\ \text{kg}\cdot\text{m}^{-3}(60^2 - 15^2)\ \text{m}^2\cdot\text{s}^{-2}$$
$$= 2059\ \text{Pa gauge (if velocities uniform)}$$

$$2059 \times \frac{\pi}{4}(0.6)^2\ \text{N} - F = \varrho Q(u_2 \cos 45° - u_1)$$
$$= 1.22\ \text{kg}\cdot\text{m}^{-3} \times \frac{\pi}{4}(0.6\ \text{m})^2 15\ \text{m}\cdot\text{s}^{-1}\left(\frac{60}{\sqrt{2}} - 15\right)\ \text{m}\cdot\text{s}^{-1}$$

whence $F =$ **440 N**

4.4

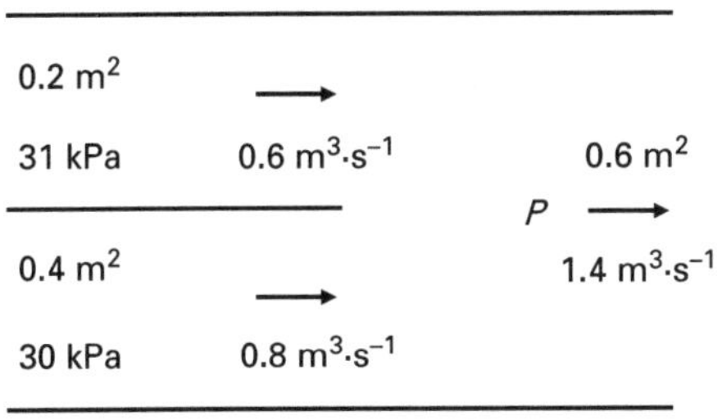

Assuming uniform conditions over each cross-section considered,

$(31 \times 10^3\ \text{Pa})(0.2\ \text{m}^2)$

$\quad + (30 \times 10^3\ \text{Pa})$

$\quad \times (0.4\ \text{m}^2) - p(0.6\ \text{m}^2)$

$$= 1000\ \text{kg} \cdot \text{m}^{-3} \left[1.4\left(\frac{1.4}{0.6}\right) - 0.6\left(\frac{0.6}{0.2}\right) - 0.8\left(\frac{0.8}{0.4}\right)\right] \text{m}^4 \cdot \text{s}^{-2}$$

$$= 1000[3.2\dot{6} - 1.8 - 1.6]\ \text{N} = -133.\dot{3}\ \text{N}$$

$$\therefore p = \frac{1}{0.6}[6200 + 12\,000 + 133.\dot{3}]\ \text{Pa} = 30.\dot{5}\ \text{kPa}$$

$$\text{Energy/mass of stream A} = \frac{p_A}{\varrho} + \frac{u_A^2}{2} = \left(31 + \frac{3^2}{2}\right) \text{m}^2 \cdot \text{s}^{-2}$$

$$= 35.5\ \text{m}^2 \cdot \text{s}^{-2}$$

$$\text{Energy/mass of stream B} = \left(30 + \frac{2^2}{2}\right) \text{m}^2 \cdot \text{s}^{-2}$$

$$= 32.0\ \text{m}^2 \cdot \text{s}^{-2}$$

$$\text{Energy/mass of stream C} = \left\{30.\dot{5} + \frac{1}{2}\left(\frac{1.4}{0.6}\right)^2\right\} \text{m}^2 \cdot \text{s}^{-2}$$

$$= 33.2\dot{7}\ \text{m}^2 \cdot \text{s}^{-2}$$

$\therefore$ Stream A loses $\mathbf{2.\dot{2}\ m^2 \cdot s^{-2}}$ Stream B loses $\mathbf{-1.2\dot{7}\ m^2 \cdot s^{-2}}$

Total loss by stream A $= 600\ \text{kg} \cdot \text{s}^{-1} \times 2.\dot{2}\ \text{m}^2 \cdot \text{s}^{-2} = 1333.\dot{3}\ \text{W}$

Total loss by stream B $= 800\ \text{kg} \cdot \text{s}^{-1} \times (-1.2\dot{7})\ \text{m}^2 \cdot \text{s}^{-2}$

$= -1022.\dot{2}\ \text{W}$

Difference $= (1333.\dot{3} - 1022.\dot{2})\ \text{W} = 311.\dot{1}\ \text{W}$

4.5 For steady flow, coordinate axes must move with boat.
For uniform conditions at inlet and outlet $F = \varrho Q(u - c)$
If static pressures fore and aft are equal, this is total propulsive force.

$\therefore$ Useful output power $= Fc = \varrho Qc(u - c)$

Power of spent jet (wasted) $= \frac{1}{2}\varrho Q(u - c)^2$

$$\therefore \text{Total power} = \varrho Q\left\{c(u-c)+\frac{1}{2}(u-c)^2\right\}$$
$$= \varrho Q(u-c)\frac{1}{2}(u+c)$$
$$\therefore \eta = \frac{\varrho Qc(u-c)}{\frac{1}{2}\varrho Q(u-c)(u+c)} = \frac{\mathbf{2c}}{\boldsymbol{u+c}}$$

With assumptions as before

$$1500\ \text{N} = 1000\ \text{kg}\cdot\text{m}^{-3} \times Q(17.5-9.3)\ \text{m}\cdot\text{s}^{-1}$$

$$\therefore Q = \frac{1.5}{8.2}\ \text{m}^3\cdot\text{s}^{-1};\quad A = \frac{1.5}{8.2\times 17.5}\ \text{m}^2 = \mathbf{0.01045\ m^2}$$

$$\text{Head to be supplied by pump} = \frac{17.5^2-9.3^2}{19.62}\ \text{m} = 11.2\ \text{m}$$

$$\therefore \text{Power} = \frac{1.5}{8.2}\ \text{m}^3\cdot\text{s}^{-1}\times 1000\ \text{kg}\cdot\text{m}^{-3}\times 9.81\ \text{N}\cdot\text{kg}^{-1}\times 11.2\ \text{m}$$
$$= 20.1\ \text{kW}$$

$$\therefore \text{Engine power} = \frac{20.1}{0.65}\ \text{kW} = \mathbf{30.92\ kW}$$

4.6 $$\dot{m}u_r = M\frac{dv}{dt} - \Sigma F = 0.086\ \text{kg}\times 15\ \text{m}\cdot\text{s}^{-2} - (-0.086\times 9.81\ \text{N})$$
$$= 2.134\ \text{N} = \dot{m}\times\frac{\dot{m}}{\varrho A}$$
$$\therefore \dot{m}^2 = 2.134\ \text{N}\times 1.29\ \text{kg}\cdot\text{m}^{-3}\frac{\pi}{4}(0.006\ \text{m})^2$$

whence $\dot{m} = 0.00882\ \text{kg}\cdot\text{s}^{-1} = \mathbf{8.82\ g\cdot s^{-1}}$

4.7 $$\Sigma F = -\left(1450\ \text{N}\cdot\text{s}\cdot\text{m}^{-1}\right)v = M\frac{dv}{dt} - \dot{m}u_r$$
$$= \left\{2500\ \text{kg} + x - \left(90\ \text{kg}\cdot\text{s}^{-1}\right)t\right\}\frac{dv}{dt} - 90\ \text{kg}\cdot\text{s}^{-1}\times 2600\ \text{m}\cdot\text{s}^{-1}$$
$$\therefore \frac{dv}{234\,000-1450v} = \frac{dt}{2500+x-90t}$$
$$\frac{1}{1450}\ln\left(\frac{234\,000-1450v}{234\,000}\right) = \frac{1}{90}\ln\left(\frac{2500+x-90t}{2500+x}\right)$$

since $v=0$ where $t=0$

v is max when all fuel burnt, that is, when $t = x/90$ s

$$\therefore \frac{90}{1450}\ln\left(\frac{234-1.45\times 150}{234}\right) = \ln\left(\frac{2500}{2500+x}\right)$$

whence $x = \mathbf{447\ kg}$

4.8 $$u_4 = 2u_2 - u_1 = \frac{2\times 4.25}{(\pi/4)(0.6)^2}\ \text{m}\cdot\text{s}^{-1} - 12\ \text{m}\cdot\text{s}^{-1} = 18.06\ \text{m}\cdot\text{s}^{-1}$$

$$F = \varrho Q(u_4 - u_1) = 1000 \times 4.25(18.06 - 12) \text{ N} = \mathbf{25.77\ kN}$$

$$\eta = \frac{2u_1}{u_1 + u_4} = \frac{u_1}{u_2} = \frac{12}{15.03} = \mathbf{79.8\%}$$

$$P = \varrho Q \frac{u_4^2 - u_1^2}{2} = \frac{\varrho Q}{2}\left(4u_2^2 - 4u_1u_2\right)$$

$$= \frac{1000 \times 4.25}{2} 4 \times 15.03(15.03 - 12) \text{ W} = \mathbf{387\ kW}$$

4.9 $u_1 = 288 \text{ km} \cdot \text{h}^{-1} = 80 \text{ m} \cdot \text{s}^{-1} \quad u_2 = u_1/0.9$

$$F = \varrho Q(u_4 - u_1) = \varrho Q(2u_2 - u_1 - u_1) = 2\varrho Q(u_2 - u_1)$$

$$= 2\varrho A \frac{u_1}{0.9}\left(\frac{u_1}{0.9} - u_1\right) = \frac{0.2}{0.81}\varrho A u_1^2 = 10\,300 \text{ N}$$

$$\therefore A = \frac{10\,300 \times 0.81}{0.2 \times 1.2 \times 80^2} \text{ m}^2 = 5.43 \text{ m}^2 \quad \therefore d = \mathbf{2.63\ m}$$

$$P_{\text{ideal}} = \frac{1}{2}\varrho Q\left(u_4^2 - u_1^2\right) = 2\varrho Q\left(u_2^2 - u_1u_2\right) = Fu_2 = \frac{Fu_1}{0.9}$$

$$= \frac{10\,300 \times 80}{0.9} \text{ W} = \mathbf{916\ kW}$$

4.10 $$\eta = 0.5 = \frac{(14 \text{ m} \cdot \text{s}^{-1} + u_4)(14^2 \text{ m}^2 \cdot \text{s}^{-2} - u_4^2)}{2 \times 14^3 \text{ m}^3 \cdot \text{s}^{-3}}$$

$$\therefore u_4 = 7\left(\sqrt{5} - 1\right) \text{ m} \cdot \text{s}^{-1} = 8.65 \text{ m} \cdot \text{s}^{-1}$$

(Negative root and $u_4 = 0$ are rejected.)

$$F = \varrho Q(u_1 - u_4) = \varrho A \frac{u_1 + u_4}{2}(u_1 - u_4) = \frac{1}{2}\varrho A\left(u_1^2 - u_4^2\right)$$

$$= \frac{1}{2} \times 1.235 \frac{\pi}{4} 12^2(14^2 - 8.65^2) \text{ N} = \mathbf{8460\ N}$$

$$u_2 = \frac{u_1 + u_4}{2} = \frac{14 + 8.65}{2} \text{ m} \cdot \text{s}^{-1} = \mathbf{11.33\ m \cdot s^{-1}}$$

$$p_2 - p_1 = \frac{1}{2}\varrho\left(u_1^2 - u_2^2\right) = \frac{1}{2}1.235(14^2 - 11.33^2) \text{ Pa}$$

$$= \mathbf{41.8\ Pa} \text{ (gauge)}$$

$$p_3 - p_4 = \frac{1}{2}\varrho\left(u_4^2 - u_3^2\right) = \frac{1}{2}1.235(8.65^2 - 11.33^2) \text{ Pa}$$

$$= \mathbf{-32.99\ Pa} = p_3 \text{ (gauge)}$$

$$P = \frac{1}{2}\varrho Q\left(u_1^2 - u_4^2\right) = \frac{1}{2}\varrho A u_2\left(u_1^2 - u_4^2\right)$$

$$= \frac{1}{2} \times 1.235 \frac{\pi}{4} 12^2 \times 11.33(14^2 - 8.65^2) \text{ W} = \mathbf{95.8\ kW}$$

Chapter 5

5.1 *Re* same in both cases.

$$\therefore\ u_{\text{w}} = u_{\text{a}}\frac{\mu_{\text{w}}}{\mu_{\text{a}}}\frac{\varrho_{\text{a}}}{\varrho_{\text{w}}}$$

$$= 21.5\ \text{m}\cdot\text{s}^{-1}\frac{1.12\times 10^{-3}}{1.8\times 10^{-5}}\frac{1.225}{1000}$$

$$= 1.639\ \text{m}\cdot\text{s}^{-1}$$

$$\therefore Q_{\text{w}} = \frac{\pi}{4}(0.04\ \text{m})^2 1.639\ \text{m}\cdot\text{s}^{-1} = 2.060\times 10^{-3}\ m^3\cdot s^{-1}$$

$$= \mathbf{2.06\ L\cdot s^{-1}}$$

$$\frac{(\Delta p^*/l)_{\text{air}}}{(\Delta p^*/l)_{\text{water}}} = \frac{\varrho_{\text{a}}u_{\text{a}}^2}{\varrho_{\text{w}}u_{\text{w}}^2} = \frac{1.225\times 21.5^2}{1000\times 1.639^2} = 0.210 = \frac{\mathbf{1}}{\mathbf{4.74}}$$

5.2 $$Q = g^{1/2}h^{5/2}\phi\left(\frac{g^{1/2}h^{3/2}\varrho}{\mu}, \frac{gh^2\varrho}{\gamma}, \theta\right)$$

For *dynamic similarity*, $g^{1/2}h^{3/2}\varrho/\mu$ must be the same.

Assume effect of γ negligible. $\therefore \dfrac{h_{\text{liquid}}}{h_{\text{water}}} = \left(\dfrac{\nu_{\text{liquid}}}{\nu_{\text{water}}}\right)^{2/3} = 4$

Also $\dfrac{Q_1}{h_1^{5/2}} = \dfrac{Q_{\text{w}}}{h_{\text{w}}^{5/2}}$

$$\therefore\ Q_{\text{w}} = \left(\frac{h_{\text{w}}}{h_1}\right)^{5/2} Q_1 = \frac{Q_1}{32} = 0.000625\ \text{m}^3\cdot\text{s}^{-1}$$

$$\therefore\ 0.000625 = 0.762h_{\text{w}}^{2.47} \quad \text{whence } h_{\text{w}} = 0.0563\ \text{m}$$

$$\therefore\ h_1 = 4h_{\text{w}} = \mathbf{225.2\ mm}$$

5.3 Equate $\varrho\omega D^2/\mu$.

$$\therefore \omega_2 = \frac{\mu_2\varrho_1\omega_1 D_1^2}{\mu_1\varrho_2 D_2^2}$$

$$= \frac{1.86 \times 10^{-5} \times 1000 \times 144.5 \times 0.225^2}{1.01 \times 10^{-3} \times 1.20 \times 0.675^2}\ \text{rad}\cdot\text{s}^{-1}$$

$$= \mathbf{246.4\ rad \cdot s^{-1}}$$

Equate $P/\varrho\omega^3 D^5$. $\therefore\ P_2 = \dfrac{\varrho_2\omega_2^3 D_2^5 P_1}{\varrho_1\omega_1^3 D_1^5}$ and

$$T_2 = \frac{P_2}{\omega_2} = \frac{\varrho_2\omega_2^3 D_2^5 T_1\omega_1}{\varrho_1\omega_1^3 D_1^5\omega_2}$$

$$= \frac{\varrho_2}{\varrho_1}\left(\frac{\omega_2}{\omega_1}\right)^2\left(\frac{D_2}{D_1}\right)^5 T_1$$

$$= \frac{1.20}{1000}\left(\frac{246.4}{144.5}\right)^2\left(\frac{0.675}{0.225}\right)^5 1.1\ \text{N}\cdot\text{m}$$

$$= \mathbf{0.933\ N \cdot m}$$

5.4 $$\frac{Q}{\omega D^3} = \phi\left(\frac{\omega d^2}{\nu}, \frac{d}{D}\right)$$

d/D is same in each case; so is $\omega d^2/\nu$

$\therefore$ Systems are dynamically similar.

$$\therefore\ Q_{\text{air}} = \frac{\omega_{\text{air}} D_{\text{air}}^3 Q_{\text{water}}}{\omega_{\text{water}} D_{\text{water}}^3} = \frac{10.9}{20.7}\left(\frac{750}{150}\right)^3 42.5 \times 10^{-3}\ \text{m}^3\cdot\text{s}^{-1}$$

$$= \mathbf{2.797\ m^3 \cdot s^{-1}}$$

5.5 Equate Re: $\dfrac{1.20 \times l \times 60\ \text{m}\cdot\text{s}^{-1}}{1.86 \times 10^{-5}} = \dfrac{1000(l/2)u_{\text{w}}}{1.01 \times 10^{-3}}$

whence $u_{\text{w}} = \mathbf{7.82\ m \cdot s^{-1}}$

Since $\dfrac{F}{l^2\varrho u^2} = \phi(Re)$, equate $\dfrac{F}{l^2\varrho u^2}$: $\dfrac{F_{\text{air}}}{l^2 \times 1.20 \times 60^2}$

$= \dfrac{1140\ \text{N}}{(l/2)^2 1000 \times 7.82^2}$ whence $F_{\text{air}} = \mathbf{322\ N}$

5.6 $\dfrac{T}{\varrho u^2 l^3} = \phi(Re)$ or equivalent is readily proved.

Equate Re: $\dfrac{u_{\text{m}}(l/4)\varrho_{\text{m}}}{\mu_{\text{m}}} = \dfrac{(3.5\ \text{m}\cdot\text{s}^{-1})l(1.025\varrho_{\text{m}})}{1.07\mu_{\text{m}}}$

$$\therefore\ u_{\mathrm{m}} = \mathbf{13.41\ m\cdot s^{-1}}$$

$$\text{Equate } \frac{T}{\varrho u^2 l^3} : T_{\mathrm{p}} = \frac{\varrho_{\mathrm{p}} u_{\mathrm{p}}^2 l_{\mathrm{p}}^3 T_{\mathrm{m}}}{\varrho_{\mathrm{m}} u_{\mathrm{m}}^2 l_{\mathrm{m}}^3} = 1.025 \left(\frac{3.5}{13.41}\right)^2 4^3 \times 20.6\ \mathrm{N\cdot m}$$

$$= \mathbf{92.0\ N\cdot m}$$

5.7 Fr the same. $\therefore$ Velocity $\propto (gl)^{1/2}$

$\therefore$ Flow $\propto l^2 (gl)^{1/2} \propto l^{5/2}$

5.8 $C_D = \phi(Re)$.

$$\text{Equate } Re : u_{\mathrm{water}} = \frac{d_{\mathrm{air}}}{d_{\mathrm{water}}} \frac{\varrho_{\mathrm{air}}}{\varrho_{\mathrm{water}}} \frac{\mu_{\mathrm{water}}}{\mu_{\mathrm{air}}} u_{\mathrm{air}}$$

$$= 2 \times \frac{1.2}{1000} \times \frac{1.01 \times 10^{-3}}{1.86 \times 10^{-5}} \times 30\ \mathrm{m\cdot s^{-1}}$$

$$= \mathbf{3.91\ m\cdot s^{-1}}$$

$$\therefore\ C_D \text{ equal } \therefore\ \mathrm{Drag}_{\mathrm{air}} = \frac{(\mathrm{Area})_{\mathrm{air}} \left(\frac{1}{2}\varrho u^2\right)_{\mathrm{air}}}{(\mathrm{Area})_{\mathrm{water}} \left(\frac{1}{2}\varrho u^2\right)_{\mathrm{water}}} F_{\mathrm{water}}$$

$$= 4 \times \frac{1.2}{1000} \times \left(\frac{30}{3.91}\right)^2 152\ \mathrm{N} = \mathbf{42.96\ N}$$

5.9 $$\text{Equate } Fr : \frac{l_{\mathrm{m}}}{l_{\mathrm{p}}} = \frac{u_{\mathrm{m}}^2}{u_{\mathrm{p}}^2} = \left(\frac{Q_{\mathrm{m}}}{Q_{\mathrm{p}}}\right)^2 \left(\frac{l_{\mathrm{p}}}{l_{\mathrm{m}}}\right)^4$$

$$\therefore\ \frac{l_{\mathrm{p}}}{l_{\mathrm{m}}} = \left(\frac{Q_{\mathrm{p}}}{Q_{\mathrm{m}}}\right)^{2/5} = \left(\frac{120}{0.75}\right)^{2/5} = \mathbf{7.61}$$

$$\text{Since } \frac{F}{l^2 \varrho u^2} = \phi(Fr), \quad \text{equate } \frac{F}{l^2 \varrho u^2}$$

$$\therefore\ F_{\mathrm{p}} = \frac{l_{\mathrm{p}}^2 u_{\mathrm{p}}^2}{l_{\mathrm{m}}^2 u_{\mathrm{m}}^2} F_{\mathrm{m}} = \left(\frac{Q_{\mathrm{p}}}{Q_{\mathrm{m}}}\right)^2 \left(\frac{l_{\mathrm{m}}}{l_{\mathrm{p}}}\right)^2 F_{\mathrm{m}} = \left(\frac{Q_{\mathrm{p}}}{Q_{\mathrm{m}}}\right)^2 \left(\frac{Q_{\mathrm{m}}}{Q_{\mathrm{p}}}\right)^{4/5} F_{\mathrm{m}}$$

$$= \left(\frac{Q_{\mathrm{p}}}{Q_{\mathrm{m}}}\right)^{6/5} F_{\mathrm{m}} = \left(\frac{120}{0.75}\right)^{6/5} 2.8\ \mathrm{N} = \mathbf{1236\ N}$$

5.10 $$\text{Equate Mach no} : u_{\mathrm{m}} = u_{\mathrm{p}} \left(\frac{\gamma R T_{\mathrm{m}}}{\gamma R T_{\mathrm{p}}}\right)^{1/2}$$

$$= 400\ \mathrm{m\cdot s^{-1}} \left(\frac{288.15}{228.15}\right)^{1/2}$$

$$= \mathbf{450\ m\cdot s^{-1}}$$

$$\text{Equate } Re: \frac{u_m}{u_p} = \left(\frac{T_m}{T_p}\right)^{1/2} = \frac{l_p \varrho_p \mu_m}{l_m \varrho_m \mu_p}$$

$$= \frac{l_p}{l_m}\frac{p_p}{p_m}\frac{T_m}{T_p} \times \left(\frac{T_m}{T_p}\right)^{3/2}\left(\frac{T_p + 177}{T_m + 117}\right)$$

$$\therefore \frac{p_m}{p_p} = \left(\frac{T_m}{T_p}\right)^2 \frac{l_p}{l_m}\left(\frac{T_p + 177}{T_m + 117}\right) = \left(\frac{288.15}{228.15}\right)^2 20\frac{345.15}{405.15}$$

whence $p_m = \mathbf{821\ kPa}$

5.11 $$\text{Equate } Fr: \therefore \frac{t_m}{t_p} = \frac{l_m}{l_p}\frac{u_p}{u_m} = \frac{l_m}{l_p}\left(\frac{l_p}{l_m}\right)^{1/2} = \left(\frac{l_m}{l_p}\right)^{1/2}$$

$$\therefore t_m = \frac{12.4\ \text{h}}{100^{1/2}} = \mathbf{1.24\ h}$$

5.12 Equate $\frac{p - p_v}{\frac{1}{2}\varrho u^2}$, that is, equate $\frac{p - p_v}{u^2}$, that is, equate $\frac{p - p_v}{h}$

$$\therefore (p - p_v)_m = \frac{1}{12}(p - p_v)_p$$

$$= \frac{1}{12}\left(1.013 \times 10^5 - 1000 \times 9.81 \times 7 - 1230\right)\ \text{Pa}$$

$$= 2617\ \text{Pa}$$

$$\therefore p_m = (2617 + 2340)\ \text{Pa} = 4957\ \text{Pa}$$

$$= (p_0)_m - 1000 \times 9.81 \times \frac{7}{12}\ \text{Pa}$$

$$\therefore (p_0)_m = (4957 + 5722)\ \text{Pa} = \mathbf{10.68\ kPa}$$

$$\frac{Q_m}{Q_p} = \frac{u_m}{u_p}\left(\frac{l_m}{l_p}\right)^2 = \left(\frac{l_m}{l_p}\right)^{5/2}$$

$$\therefore Q_m = \frac{7\ \text{m}^3 \cdot \text{s}^{-1}}{12^{5/2}} = \mathbf{14.03 \times 10^{-3}\ m^3 \cdot s^{-1}}$$

5.13 Equate *Fr*. Then $u_p = u_m(l_p/l_m)^{1/2} = (1.5\ \text{m} \cdot \text{s}^{-1})(100/4)^{1/2}$

$$= \mathbf{7.5\ m \cdot s^{-1}}$$

Skin friction $= kAu^n$. For model, $14.5\ \text{N} \cdot \text{m}^{-2} = k_m(3\ \text{m} \cdot \text{s}^{-1})^{1.9}$

whence $k_m = 1.798\ \text{N} \cdot \text{s}^{1.9} \cdot \text{m}^{-3.9}$

$$\therefore \text{At } 1.5\ \text{m} \cdot \text{s}^{-1} (\text{Skin friction})_m = 1.798 \times \frac{1200}{25^2} \times 1.5^{1.9}\ \text{N}$$

$$= 7.46\ \text{N}$$

$\therefore$ (Resid. resistance)$_{\mathrm{m}}$ = (15.5 – 7.46) N = 8.04 N

$$(F_{\mathrm{resid}})_{\mathrm{p}} = (F_{\mathrm{resid}})_{\mathrm{m}} \frac{\varrho_{\mathrm{p}}}{\varrho_{\mathrm{m}}} \frac{u_{\mathrm{p}}^2}{u_{\mathrm{m}}^2} \frac{l_{\mathrm{p}}^2}{l_{\mathrm{m}}^2} = (F_{\mathrm{resid}})_{\mathrm{m}} \frac{\varrho_{\mathrm{p}}}{\varrho_{\mathrm{m}}} \left(\frac{l_{\mathrm{p}}}{l_{\mathrm{m}}}\right)^3$$

$$= 8.04 \times 1.026 \times 25^3 \text{ N} = 128.9 \text{ kN}$$

$$(\text{Skin friction})_{\mathrm{p}} = k_{\mathrm{p}}(\text{Area})_{\mathrm{p}} u_{\mathrm{p}}^{1.85}$$

$$\therefore k_{\mathrm{p}} = 43 \text{ N} \cdot \text{m}^{-2} (3 \text{ m} \cdot \text{s}^{-1})^{-1.85} = 5.63 \text{ N} \cdot \text{s}^{1.85} \cdot \text{m}^{-3.85}$$

$\therefore$ At 7.5 $\text{m} \cdot \text{s}^{-1}$

$$(\text{Skin friction})_{\mathrm{p}} = 5.63 \times 1200 \times 7.5^{1.85} \text{ N}$$

$$= 281.1 \text{ kN}$$

$\therefore$ (Total resistance)$_{\mathrm{p}}$ = (128.9 + 281.1) kN = **410 kN**

Chapter 6

6.1 $$Re = \frac{ud\varrho}{\mu} = \frac{4Q\varrho}{\pi d\mu} = \frac{4 \times 0.020\ \text{m}^2 \cdot \text{s}^{-1} \times 1260\ \text{kg} \cdot \text{m}^{-3}}{\pi 0.1\ \text{m} \times 0.9\ \text{N} \cdot \text{s} \cdot \text{m}^{-2}} = \mathbf{357}$$

$$\Delta p^* = \frac{128\mu Q l}{\pi d^4} = \frac{128 \times 0.9 \times 0.020 \times 45}{\pi 0.1^4} \text{N} \cdot \text{m}^{-2} = 3.3 \times 10^5\ \text{Pa}$$

$\therefore$ p at upper end

$$= (5.9 \times 10^5 - 3.3 \times 10^5 - 1260 \times 9.81 \times 45 \sin 15°)\ \text{Pa}$$

$$= \mathbf{116\ kPa}$$

$$\overline{\tau}_0 = \frac{d}{4}\frac{\Delta p^*}{l} = \frac{0.1}{4}\frac{3.3 \times 10^5}{45}\ \text{Pa} = \mathbf{183.3\ Pa}$$

6.2 From eqn 6.19, $u = -k(y^2 - cy)$ where k = constant.

$$\frac{\text{Total rate of K.E. flux divided by width}}{\text{Total rate of mass flow divided by width}}$$

$$= \frac{\int_0^c \frac{1}{2}(\varrho u)u^2\, \text{d}y}{\int_0^c \varrho u\, \text{d}y} = \frac{\int_0^c u^3\, \text{d}y}{2\int_0^c u\, \text{d}y}$$

$$= \frac{-k^3 \int_0^c (y^6 - 3cy^5 + 3c^2y^4 - c^3y^3)\, \text{d}y}{-2k\int_0^c (y^2 - cy)\, \text{d}y} = \frac{k^2}{2}\frac{3c^4}{70}$$

$$u_\text{m} = \frac{1}{c}\int_0^c u\, \text{d}y = \frac{kc^2}{6} \quad \therefore kc^2 = 6u_\text{m}$$

$$\therefore \frac{\text{K.E.}}{\text{Mass}} = \frac{36u_\text{m}^2}{2} \times \frac{3}{70} = \mathbf{1.543}\boldsymbol{u_\text{m}^2}\mathbf{/2}$$

6.3 $$B = \frac{4Q\mu_\text{p}}{\pi R^3|\tau_y|} + \frac{4}{3} = \frac{4 \times 0.01\ \text{m}^3 \cdot \text{s}^{-1} \times 1.6\ \text{Pa} \cdot \text{s}}{\pi (0.05\ \text{m})^3 \times 120\ \text{Pa}} + \frac{4}{3} = 2.691$$

From eqn 6.15, $m(\text{finally}) = 2.674 = \tau_0/\tau_y \quad \therefore |\tau_0| = 320.9$ Pa

$$\therefore \frac{dp^*}{dx} = \frac{4\tau_0}{d} = -\frac{4 \times 320.9 \text{ Pa}}{0.1 \text{ m}} = -12\,835 \text{ Pa} \cdot \text{m}^{-1}$$

$\therefore \Delta p = 12\,835 \text{ Pa} \cdot \text{m}^{-1} \times 15 \text{ m} = \mathbf{192.5\ kPa}$

6.4 Case (a): $\text{Torque} = 2\pi r l \mu \dfrac{du}{dr} r = 2\pi r^2 l \mu \dfrac{\omega r}{t}$

$\therefore \text{Power} = T\omega = 2\pi r^3 l \mu \omega^2 / t$

Case (b): $p = 12\mu l \bar{u}/t^2$ (from eqn 6.21)

$\text{Power} = Qp = 2\pi r t \bar{u} p = 2\pi r t^3 p^2 / 12\mu l$

Equate powers: $\dfrac{2\pi r^3 l \mu \omega^2}{t} = \dfrac{2\pi r t^3 p^2}{12\mu l}$

whence $p = \dfrac{\mathbf{2\mu l r \omega \sqrt{3}}}{\mathbf{c^2}}$

6.5 From eqn 6.20 $\dfrac{dp}{dr} = -\dfrac{12\mu Q}{2\pi r c^3}$

$$\therefore \int_p^0 dp = \int_r^{R_2} -\frac{6\mu Q}{\pi r c^3} dr = \frac{6\mu Q}{\pi c^3} \ln \frac{r}{R_2} = -p \qquad (1)$$

$\therefore$ Total lifting force between radii R_1 and R_2 is

$$-\int_{R_1}^{R_2} \frac{6\mu Q}{\pi c^3} \ln \frac{r}{R_2} 2\pi r \, dr = \frac{6\mu Q}{c^3} \left[\frac{R_2^2 - R_1^2}{2} - R_1^2 \ln \frac{R_2}{R_1} \right]$$

But from eqn (1) above

$$\frac{6\mu Q}{c^3} = \frac{\pi p_1}{\ln(R_2/R_1)}$$

and there is additional lifting force $p_1 \pi R_1^2$ between radii 0 and R_1

$$\therefore \text{Total lifting force} = \frac{\pi p_1}{\ln(R_2/R_1)} \left[\frac{R_2^2 - R_1^2}{2} - R_1^2 \ln \frac{R_2}{R_1} \right] + p_1 \pi R_1^2$$

$$= \frac{\boldsymbol{\pi p_1 \left(R_2^2 - R_1^2\right)}}{\mathbf{2 \ln(R_2/R_1)}}$$

$$c^3 = \frac{6\mu Q \ln(R_2/R_1)}{\pi p_1} = \frac{6 \times 0.08 \text{ Pa} \cdot \text{s} \times 0.85 \times 10^{-3} \text{ m}^3 \cdot \text{s}^{-1} \ln 4}{\pi 5.5 \times 10^5 \text{ Pa}}$$

$= 3.273 \times 10^{-10} \text{ m}^3$

$\therefore c = \mathbf{0.689\ mm}$

6.6 $$\frac{dh}{dt}2\pi r\delta r = (\overline{u}+\delta\overline{u})2\pi(r+\delta r)h - \overline{u}2\pi rh = 2\pi h(r\delta\overline{u}+\overline{u}\delta r)$$

$$\therefore r\frac{dh}{dt} = h\frac{d}{dr}(\overline{u}r)$$

$$\therefore \frac{r^2}{2}\frac{dh}{dt} = h\overline{u}r + \text{const.}$$ Const. = 0 to satisfy conditions at $r = 0$

$$\therefore \overline{u} = \frac{r}{2h}\frac{dh}{dt}$$

From eqn 6.20, $\dfrac{dp}{dr} = -\dfrac{12\mu\overline{u}}{h^2} = -\dfrac{6\mu r}{h^3}\dfrac{dh}{dt}$

$$\int_p^0 dp = -p = -\frac{6\mu}{h^3}\frac{dh}{dt}\int_r^a r\,dr = -\frac{3\mu}{h^3}\frac{dh}{dt}(a^2 - r^2)$$

$$F = \int_0^a p\,2\pi r\,dr = \frac{3\mu}{h^3}\frac{dh}{dt}2\pi\int_0^a (a^2 r - r^3)dr$$

$$= \frac{6\pi\mu}{h^3}\frac{dh}{dt}\left[\frac{a^4}{2} - \frac{a^4}{4}\right]$$

$$= \boldsymbol{\frac{3\pi\mu a^4}{2h^3}\frac{dh}{dt}}$$

6.7 Weight – buoyancy of piston

$$= \left(9 \times 9.81 - \frac{\pi}{4}0.113^2 \times 0.15 \times 900 \times 9.81\right)\text{N}$$

$$= 75.0\text{ N}$$

Using mean diameter 114 mm, eqn 6.27 gives

$$V_p = \frac{4 \times 75.0\text{ N}}{3\pi 0.12\text{ N}\cdot\text{s}\cdot\text{m}^{-2} \times 0.15\text{ m}(114/1)^3} = 1.194 \times 10^{-3}\text{ m}\cdot\text{s}^{-1}$$

$$\therefore t = \frac{0.075\text{ m}}{1.194 \times 10^{-3}\text{ m}\cdot\text{s}^{-1}} = \mathbf{62.8\ s}$$

6.8 $$Re = \frac{ud}{\nu} = \frac{d}{\nu}\frac{d^2 g}{18\mu}(\varrho_{\text{dust}} - \varrho_{\text{air}})$$

Max. d^3 is when $Re = 0.1$. Then $d^3 = \dfrac{18\varrho\nu^2 0.1}{g(\varrho_{\text{dust}} - \varrho_{\text{air}})}$

$$= \frac{18 \times 1.225\text{ kg}\cdot\text{m}^{-3} \times 14.9^2 \times 10^{-12}\text{ m}^4\cdot\text{s}^{-2} \times 0.1}{9.81\text{ m}\cdot\text{s}^{-2} \times (2500 - 1.225)\text{ kg}\cdot\text{m}^{-3}}$$

whence $d = 2.713 \times 10^{-5}\text{ m} = \mathbf{27.13\ \mu m}$

$$u = Re\frac{\nu}{d} = \frac{0.1 \times 14.9 \times 10^{-6}\text{ m}^2\cdot\text{s}^{-1}}{2.713 \times 10^{-5}\text{ m}} = \mathbf{0.0549\ m\cdot s^{-1}}$$

6.9
$$\varrho_s = \frac{13.7 \times 10^{-6}\ \text{kg}}{\pi/6 \times 1.5^3 \times 10^{-9}\ \text{m}^3} = 7750\ \text{kg}\cdot\text{m}^{-3}$$

$$\mu = \frac{d^2 g(\varrho_s - \varrho)}{18u}$$

$$= \frac{1.5^2 \times 10^{-6}\ \text{m}^2 \times 9.81\ \text{N}\cdot\text{kg}^{-1} \times (7750 - 950)\ \text{kg}\cdot\text{m}^{-3}}{18 \times 0.5\ \text{m}/56\ \text{s}}$$

$$= \mathbf{0.934\ Pa\cdot s}$$

$$Re = \frac{ud\varrho}{\mu} = \frac{0.5}{56} \times 0.0015 \times \frac{950}{0.934} = \mathbf{0.01362} \quad \text{which is} <0.1$$

6.10 Difference of depths = 50 mm.

Difference of torques = 0.36 N · m

$$\mu = \frac{T(a^2 - b^2)}{4\pi h a^2 b^2 \Omega} \quad \text{[eqn 6.32]}$$

$$= \frac{0.36\ \text{N}\cdot\text{m}(0.0505^2 - 0.05^2)\ \text{m}^2}{4\pi \times 0.05\ \text{m} \times 0.0505^2\ \text{m}^2 \times 0.05^2\ \text{m}^2 \times 30\ \text{s}^{-1}}$$

$$= \mathbf{0.1505\ Pa\cdot s}$$

6.11
$$\delta = \frac{0.075 - 0.025}{150} = \frac{0.075\ \text{mm}}{a} \quad \therefore a = 225\ \text{mm}$$

$$\mu = \frac{T\delta^2}{6V}\left\{\ln\left(\frac{a}{a-l}\right) - \frac{2l}{2a-l}\right\}^{-1}$$

$$= \frac{5 \times 10^5\ \text{N}\cdot\text{m}^{-1} \times (1/3000)^2}{6 \times 1.5\ \text{m}\cdot\text{s}^{-1}}\left\{\ln\frac{225}{75} - \frac{300}{300}\right\}^{-1}$$

$$= \mathbf{0.0626\ Pa\cdot s}$$

$$\text{Drag/width} = \frac{T\delta}{2} + \frac{\mu V}{\delta}\ln\frac{h_1}{h_2}$$

$$= \left\{\frac{5 \times 10^5}{2 \times 3000} + \frac{0.0626 \times 1.5}{1/3000}\ln 3\right\}\ \text{N}$$

$$= 392.8\ N$$

$$\therefore \text{Power} = 392.8\ \text{N} \times 1.5\ \text{m}\cdot\text{s}^{-1} = \mathbf{589\ W}$$

From eqn 6.36
$$p = \frac{6\mu V x(l-x)}{\delta^2 (a-x)^2 (2a-l)}$$

$$\therefore \frac{dp}{dx} = \frac{6\mu V}{\delta^2(2a-l)}\left\{\frac{(a-x)^2(l-2x) + x(l-x)2(a-x)}{(a-x)^4}\right\} = 0$$

when $x = \dfrac{al}{2a - l}$

$$\therefore p_{\max} = \frac{3\mu V l^2}{2\delta^2 a(a-l)(2a-l)} = \mathbf{5.63 \times 10^6\ Pa}$$

From eqn 6.40

$$x_p = \frac{a(3a-2l)\ln[a/(a-l)] - l(3a - l/2)}{(2a-l)\ln\left(\dfrac{a}{a-l}\right) - 2l} = \mathbf{91.1\ mm}$$

6.12 From eqn 6.35 $\dfrac{1}{12\mu}\dfrac{dp}{dx} = \dfrac{V}{2h^2} - \dfrac{q}{h^3}$

$$= \frac{V}{2h_1^2}\exp\frac{2x}{l} - \frac{q}{h_1^3}\exp\frac{3x}{l}$$

$$\therefore \frac{p}{12\mu} = \frac{Vl}{4h_1^2}\exp\frac{2x}{l} - \frac{ql}{3h_1^3}\exp\frac{3x}{l} + C$$

where $C =$ constant. $p = 0$ at $x = 0$ and $x = l$.

$$\therefore 0 = \frac{Vl}{4h_1^2} - \frac{ql}{3h_1^3} + C \quad \text{and} \quad 0 = \frac{Vl}{4h_1^2}e^2 - \frac{ql}{3h_1^3}e^3 + C$$

Eliminating C gives $\dfrac{Vl}{4h_1^2}(e^2 - 1) = \dfrac{ql}{3h_1^3}(e^3 - 1)$

If max p occurs when $x = x_m$ and $h = h_m$
then $V/2 = q/h_m$ (from eqn 6.35)

$$\therefore \frac{2}{h_m} = \frac{2}{h_1}\exp\frac{x_m}{l} = \frac{V}{q} = \frac{4}{3h_1}\left(\frac{e^3-1}{e^2-1}\right)$$

$$\therefore \frac{x_m}{l} = \ln\left\{\frac{2}{3}\left(\frac{e^3-1}{e^2-1}\right)\right\} = \mathbf{0.689}$$

6.13 (a) $\mu = \dfrac{Fc^2(1-\varepsilon^2)^{1/2}(2+\varepsilon^2)}{12\pi\Omega R^3\varepsilon}$

$$= \frac{(20 \times 10^3\ \text{N}/0.06\ \text{m})(0.001)^2 0.8 \times 2.36}{12\pi \times 20\pi\ \text{rad}\cdot\text{s}^{-1} \times 0.025\ \text{m} \times 0.6}$$

$$= \mathbf{0.01771\ Pa \cdot s}$$

$$\text{Power} = \text{Torque} \times \Omega = \frac{4\pi\mu\Omega^2 R^3(1+2\varepsilon^2) \times \text{Length}}{c(1-\varepsilon^2)^{1/2}(2+\varepsilon^2)}$$

$$= \frac{4\pi 0.01771\ \text{Pa}\cdot\text{s}(20\pi\ \text{rad}\cdot\text{s}^{-1})^2(0.025\ \text{m})^2 1.72 \times 0.06\ \text{m}}{0.001 \times 0.8 \times 2.36}$$

$$= \mathbf{30.02\ W}$$

Supply hole ideally at position of min pressure,
that is, where $h = h_0$ that is, $c(1+\varepsilon\cos\theta) = c(1-\varepsilon^2)/(1+\varepsilon^2/2)$

$$\therefore \cos\theta = -\frac{3\varepsilon}{2+\varepsilon^2} = -\frac{1.8}{2.36} \qquad \therefore \theta = 180^\circ \pm 40.3^\circ$$

$\therefore$ Min pressure is at 220.3° from OC, that is, **130.3° from load line**

$$\text{(b)} \quad \mu = \text{Force} \times \frac{4c^2(1-\varepsilon^2)}{\pi\Omega RL^3\varepsilon}\left\{\left(\frac{16}{\pi^2}-1\right)\varepsilon^2+1\right\}^{-1/2}$$

$$= 20 \times 10^3\ \text{N} \times \frac{4 \times (0.001)^2 0.025\ \text{m}(0.64)^2}{\pi(20\pi\ \text{rad}\cdot\text{s}^{-1})(0.06\ \text{m})^3 0.6}$$

$$\times \{0.621 \times 0.6^2 + 1\}^{-1/2}$$

$$= \mathbf{0.02895\ Pa\cdot s}$$

$$\tan\psi = \frac{\pi(1-\varepsilon^2)^{1/2}}{4\varepsilon} = \frac{\pi 0.8}{4 \times 0.6} = \frac{\pi}{3} \qquad \therefore \psi = 46.32^\circ$$

$\therefore$ 130.3° from load line is about **84° from OC.**

Chapter 7

7.1 $$u = \frac{0.045\ \text{m}^3 \cdot \text{s}^{-1}}{(\pi/4)\,(0.15\ \text{m})^2} = 2.546\ \text{m} \cdot \text{s}^{-1}$$

$$Re = \frac{2.546\ \text{m} \cdot \text{s}^{-1} \times 0.15\ \text{m} \times 1830\ \text{kg} \cdot \text{m}^{-3}}{0.04\ \text{Pa} \cdot \text{s}} = 17\,480$$

$\therefore$ Turbulent flow.

$$\therefore\ f = 0.0014(1 + 100\ Re^{-1/3}) = 0.00679$$

$$h_f = \frac{u^2}{2g}\left(\frac{4fl}{d} + 1.5\right) = \frac{2.546^2}{19.62}\left(\frac{4 \times 0.00679 \times 18}{0.15} + 1.5\right)\text{m}$$
$$= 1.574\ \text{m}$$

$$\therefore\ \text{Power} = 0.045\ \text{m}^3 \cdot \text{s}^{-1} \times 1830\ \text{kg} \cdot \text{m}^{-3}$$
$$\times\ 9.81\ \text{N} \cdot \text{kg}^{-1}(6 + 1.574)\ \text{m}$$
$$= \mathbf{6120\ W}$$

7.2 $$\tau_0 \text{ for liquid} = \frac{1}{2}\varrho u^2 f = \frac{1}{2}900\ \text{kg} \cdot \text{m}^{-3}(2.5\ \text{m} \cdot \text{s}^{-1})^2 0.01$$
$$= 28.125\ \text{Pa}$$

$$\tau_0 \text{ for water} = \frac{1}{2}1000\ \text{kg} \cdot \text{m}^{-3}(2.5\ \text{m} \cdot \text{s}^{-1})^2 0.01 = 31.25\ \text{Pa}$$

$$\text{Wetted area inside one tube} = \pi 0.025 \times 3.65\ \text{m}^2$$
$$= 0.09125\pi\ \text{m}^2$$

$$\text{Wetted area outside one tube} = \pi 0.03 \times 3.65\ \text{m}^2$$
$$= 0.1095\pi\ \text{m}^2$$

$$\text{Wetted area of cylinder} = \pi 0.75 \times 3.65\ \text{m}^2$$
$$= 2.7375\pi\ \text{m}^2$$

Case I Area wetted by liquid $= 18.25\pi \text{ m}^2$

Are wetted by water $= (200 \times 0.1095\pi + 2.7375\pi) \text{ m}^2$
$= 24.64\pi \text{ m}^2$

$\therefore$ Power $= 2.5 \text{ m}\cdot\text{s}^{-1} (25.125 \times 18.25\pi + 31.25 \times 24.64\pi)\text{ N}$
$=$ **10.08 kW**

Case II Power $= 2.5 \text{ m}\cdot\text{s}^{-1}$
$\times (28.125 \times 24.64\pi + 31.25 \times 18.25\pi)\text{ N}$
$= 9.92 \text{ kW}$

$\therefore$ Saving $= 2.5 \text{ m}\cdot\text{s}^{-1}$
$\times (-28.125 \times 6.39\pi + 31.25 \times 6.39\pi)\text{ N}$
$=$ **157 W**

7.3 Jet velocity $= \sqrt{2g \times 35 \text{ m}} = 26.20 \text{ m}\cdot\text{s}^{-1}$

$$1.4 \text{ MPa} = \frac{1.4 \times 10^6}{1000 \times 9.81}\text{m} = 142.7 \text{ m head}$$

$$\therefore h_f = (142.7 - 35 - 3) \text{ m} = 104.7 \text{ m}$$

$$= \frac{4 \times 0.01 \times 450}{0.075}\frac{u^2}{2g}$$

whence u in pipe $= 2.926 \text{ m}\cdot\text{s}^{-1}$

$$d_{\text{jet}}^2 = \frac{2.926}{26.20} \times 0.075^2 \text{ m}^2 \quad \therefore d_{\text{jet}} = \mathbf{25.06\ mm}$$

$$Q = \frac{\pi}{4}(0.02506)^2 \times 26.20 \text{ m}^3\cdot\text{s}^{-1} = 0.01293 \text{ m}^3\cdot\text{s}^{-1}$$

$$\text{Power} = 0.01293 \text{ m}^3\cdot\text{s}^{-1} \times 14 \times 10^6 \text{ N}\cdot\text{m}^{-2} \times \frac{1}{0.7} = \mathbf{25.85\ kW}$$

7.4 $\bar{\tau}_0 = 200 \text{ Pa} = \Delta p^* \frac{\pi}{4} d^2/(\pi d l) = \Delta p^* d/(4l)$

$$\therefore \Delta p^* = \frac{200 \times 4 \times 60}{0.1} \text{ Pa} = 480 \text{ kPa}$$

$$\therefore p_{\text{inlet}} = \left(120 + 480 + \frac{900}{1000} \times 9.81 \times 60 \sin 10^\circ\right) \text{kPa} = \mathbf{692\ kPa}$$

$$h_f = \frac{480 \times 10^3}{900 \times 9.81}\text{m} = 54.37 \text{ m}$$

If flow is laminar $h_f = \dfrac{4l}{d}\dfrac{u^2}{2g}\dfrac{16\nu}{ud} = \dfrac{32lu\nu}{gd^2}$

$$\therefore u = \frac{gh_f d^2}{32l\nu} = 23.15 \text{ m} \cdot \text{s}^{-1}$$

$$\therefore Re = 23.15 \times 0.1/120 \times 10^{-6} = 19\,290$$

∴ Flow is *not* laminar.

If flow is turbulent $h_f = \dfrac{4l}{d}\dfrac{u^2}{2g}0.08\left(\dfrac{\nu}{ud}\right)^{1/4}$

$$\therefore u^{7/4} = \frac{gh_f d^{5/4}}{0.16 l \nu^{1/4}} = 29.85 \ (\text{m} \cdot \text{s}^{-1})^{7/4} \quad \therefore u = 6.963 \text{ m} \cdot \text{s}^{-1}$$

$$\therefore Re = 5803\text{: OK}$$

$$Q = \frac{\pi}{4}(0.1)^2\, 6.963 \text{ m}^3 \cdot \text{s}^{-1} = \mathbf{0.0547 \ m^3 \cdot s^{-1}}$$

7.5 $$h_f = \frac{4fl}{d}\frac{u_{\text{jet}}^2}{2g}\left(\frac{\text{Area of jet}}{\text{Area of hose}}\right)^2 = \frac{4f40}{d} \times 40\left(\frac{0.05}{d}\right)^4$$

$$= \frac{0.04f}{d^5} \quad [\text{m}^6]$$

$$< \frac{15}{85} \times 40\text{m} = \frac{120}{17}\text{m}$$

$$\therefore d^5 > \frac{0.04 \times 17}{120} f = \frac{17}{3000} f \quad [\text{m}^5]$$

Use *higher* value of f. Then $d^5 > \dfrac{17 \times 0.01}{3000}$

$$= \frac{17}{3} \times 10^{-5} \quad [\text{m}^5]$$

$$\therefore d > 1.415 \times 10^{-1} \text{ m} = \mathbf{141.5 \ mm, \ say \ 150 \ mm}$$

7.6 10 m water ≡ 1000 × 9.81 × 10 Pa. Now $h_f = KQ^2 l$,
where K = constant.

$$\therefore (4 + 9.81)10^4 \text{ Pa} = K(0.049)^2 900 \text{ m}^7 \cdot \text{s}^{-2}$$

$$\text{New } \Delta p = K(0.067)^2 450 \text{ m}^7 \cdot \text{s}^{-2} + K(0.049)^2 450 \text{ m}^7 \cdot \text{s}^{-2}$$

$$= \frac{13.81 \times 10^4}{2}\left\{\left(\frac{0.067}{0.049}\right)^2 + 1\right\} \text{ Pa} = 19.81 \times 10^4 \text{ Pa}$$

$$\therefore \text{Inlet pressure} = (19.81 - 9.81)10^4 \text{ Pa} = \mathbf{100 \ kPa}$$

7.7 For steady flow, head difference = Σ losses

Entry to pipe (1): $0.5u_1^2/2g$

Friction in pipes (1) and (3): $2 \times \dfrac{4fl_1}{d_1}\dfrac{u_1^2}{2g}$

where $f = 0.005\left(1 + \dfrac{25}{75}\right) = 0.00667$

$$\therefore \text{Loss} = \frac{2 \times 4 \times 0.00667 \times 7.5}{0.075}\frac{u_1^2}{2g} = \qquad 5.33u_1^2/2g$$

Enlargement, (1) → (2):

$$\text{Loss} = \frac{(u_1 - u_2)^2}{2g} = \frac{u_1^2}{2g}\left(1 - \frac{1}{16}\right)^2 = \qquad 0.88u_1^2/2g$$

$$\left[\text{since } u_2 = \frac{A_1}{A_2}u_1 = \left(\frac{d_1}{d_2}\right)^2 u_1\right]$$

Friction in pipe(2): $\dfrac{4fl_2}{d_2}\dfrac{u_2^2}{2g}$

where $f = 0.005\left(1 + \dfrac{25}{300}\right) = 0.00542$

$$\therefore \text{Loss} = \frac{4 \times 0.00542 \times 60}{0.3}\left(\frac{u_1}{16}\right)^2\frac{1}{2g} = \qquad 0.017u_1^2/2g$$

Loss at entry to (3) is $0.5u_3^2/2g =$ $0.5u_1^2/2g$

Exit loss = $u_3^2/2g =$ $1.0u_1^2/2g$

Total = $8.23u_1^2/2g$

$\therefore 1.5\text{ m} = 8.23u_1^2/2g$ whence $u_1 = 1.891\text{ m}\cdot\text{s}^{-1}$

$\therefore Q = \dfrac{\pi}{4}(0.075)^2 1.891\text{ m}^3\cdot\text{s}^{-1} = 0.00835\text{ m}^3\cdot\text{s}^{-1} = \mathbf{8.35\ L\cdot s^{-1}}$

7.8 Inlet head = $\dfrac{370 \times 10^3}{820 \times 9.81}$ m = 46.0 m; Static lift = 20 m

$$\therefore h_f = 26\text{ m} = \frac{4f185}{d19.62}\left(\frac{0.040}{(\pi/4)d^2}\right)^2\text{ m}^6$$

$$\therefore d^5 = 0.003763f\text{ m}^5$$

$$Re = \frac{0.040}{(\pi/4)d^2}\frac{d}{2.3 \times 10^{-6}}\ [\text{m}] = \frac{2.214 \times 10^4}{d}\ [\text{m}]$$

Try $f = 0.006$ Then $d = 0.1177$m; $Re = 1.886 \times 10^5$;

$$\frac{k}{d} = \frac{0.15 \times 10^{-3}}{0.1177} = 0.001274$$

$\therefore f = 0.0057$

Then $d = 0.1165$ m; $Re = 1.901 \times 10^5$; $\dfrac{k}{d} = 0.001288$

$\therefore f = 0.0057$

$\therefore d =$ **116.5 mm, say 120 mm**

7.9
$$Q \simeq -\frac{\pi}{2} d^{5/2}\left(\frac{2gh_f}{l}\right)^{1/2} \log\left\{\frac{k}{3.71d} + \frac{2.51\nu}{d^{3/2}\left(2gh_f/l\right)^{1/2}}\right\}$$

$$= -\frac{\pi}{2}(0.075)^{5/2}\left(\frac{19.62 \times 26}{185}\right)^{1/2} \log\left\{\frac{0.15 \times 10^{-3}}{3.71 \times 0.075} + \frac{2.51 \times 2.3 \times 10^{-6}}{(0.075)^{3/2}(19.62 \times 26/185)^{1/2}}\right\} \text{ m}^3 \cdot \text{s}^{-1}$$

$= 0.0127 \text{ m}^3 \cdot \text{s}^{-1} =$ **12.7 L · s⁻¹**

7.10

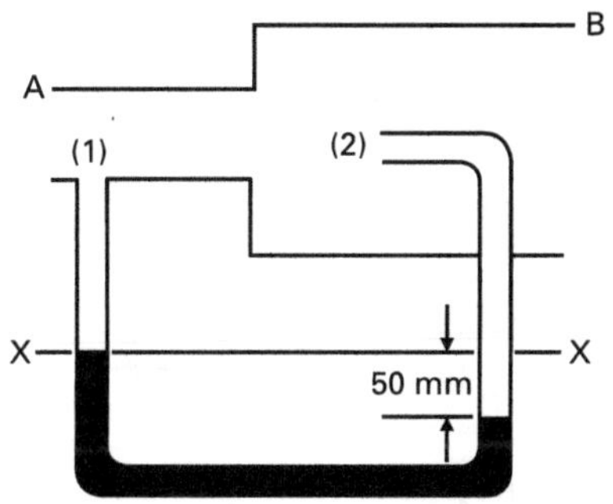

Let head difference above level XX be h metres of liquid. Then

$(h + 0.05)1200 \times 9.81$
$= 0.05 \times 13\,600 \times 9.81$
$\therefore h = 0.517$

Manometer measures $\left(\dfrac{p_2^*}{\varrho g} + \dfrac{u_2^2}{2g}\right) - \dfrac{p_1^*}{\varrho g}$

that is, $\dfrac{1}{2g}\left\{u_1^2 - (u_1 - u_2)^2\right\}$

$$= \frac{1}{2g}\left(2u_1u_2 - u_2^2\right) = \frac{Q^2}{2g}\left(\frac{2}{A_1A_2} - \frac{1}{A_2^2}\right) = h \text{ m}$$

$$\therefore Q^2 = \frac{2ghA_2}{(2/A_1) - (1/A_2)}$$

$$= \frac{19.62 \times 0.517 \times (\pi/4)(0.1)^2}{\left[2/(\pi/4)(0.05)^2\right] - \left[1/(\pi/4)(0.1)^2\right]} \text{ m}^6 \cdot \text{s}^{-2}$$

$$= 8.93 \times 10^{-5} \text{ m}^6 \cdot \text{s}^{-2}$$

$\therefore Q = 9.45 \times 10^{-3} \text{ m}^3 \cdot \text{s}^{-1}$

$\therefore$ Mass flow rate $= 9.45 \times 10^{-3} \text{ m}^3 \cdot \text{s}^{-1} \times 1200 \text{ kg} \cdot \text{m}^{-3}$

$=$ **11.34 kg · s⁻¹**

7.11 A: 

$$(h_f)_A = \frac{4fl}{d}\frac{u^2}{2g}$$

B: 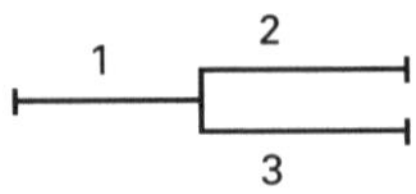

$$(h_f)_B = \frac{4fl/2}{d}\frac{u_1^2}{2g} + \frac{4fl/2}{d}\frac{u_2^2}{2g}$$

$$= \frac{2fl}{d}\frac{u_1^2}{2g}\left(1 + \frac{1}{4}\right)$$

since $u_2 = u_3 = \dfrac{u_1}{2}$

$(h_f)_B = (h_f)_A \quad \therefore \dfrac{u_1}{u} = \sqrt{\dfrac{8}{5}} = 1.265 \quad \therefore$ Increase = **26.5%**

7.12

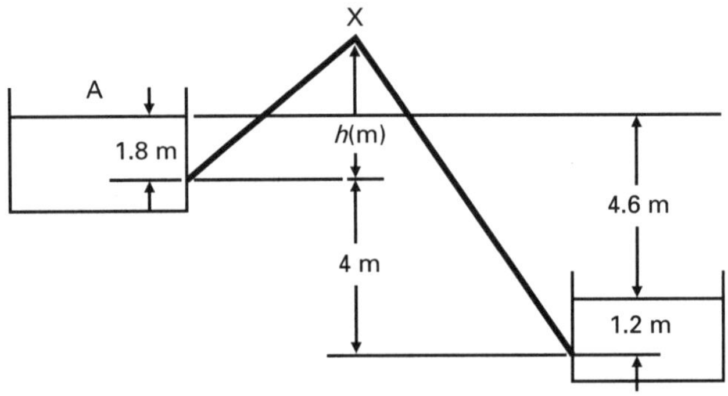

Length to X $= x = \sqrt{7.5^2 + h^2}$ [m]

$$4.6\text{ m} = \left(1.5 + \frac{4 \times 0.01 \times 36}{0.1}\right)\frac{u^2}{2g} \quad \therefore \frac{u^2}{2g} = 0.2893\text{ m}$$

$$\text{Energy/weight at X} = \frac{40 \times 10^3}{1000 \times 9.81}\text{ m} + \frac{u^2}{2g} + (h - 1.8)\text{ m}$$

= Energy/weight at A – Head losses between A and X

$$= \frac{101.3 \times 10^3}{1000 \times 9.81}\text{ m} - \left(0.5 + \frac{4fx}{d}\right)\frac{u^2}{2g}$$

$$\therefore 0.2893\left(1 + 0.5 + \frac{4 \times 0.01}{0.1}\sqrt{7.5^2 + h^2}\right) + h$$

$$= \frac{61.3 \times 10^3}{1000 \times 9.81} + 1.8 \quad \text{whence} \quad h = \mathbf{6.47}\ \mathbf{[m]}$$

7.13 $\text{Velocity at C} = \dfrac{0.0425\text{ m}^3 \cdot \text{s}^{-1}}{(\pi/4)(0.1\text{ m})^2 0.62} = 8.73\text{ m} \cdot \text{s}^{-1}$

$$\text{Velocity at B} = \frac{0.0425}{(\pi/4)(0.2)^2}\text{ m} \cdot \text{s}^{-1} = 1.353\text{ m} \cdot \text{s}^{-1}$$

Head lost C → B is $\dfrac{(8.73 - 1.353)^2}{19.62}$ m $= 2.772$ m

$= \text{Loss between A and B}$

$$\frac{p_B - p_A}{\varrho g} = \frac{u_A^2 - u_B^2}{2g} - h_1 = \frac{\{(200/150)^4 - 1\}1.353^2}{19.62}\ \text{m} - 2.772\ \text{m}$$

$$= \mathbf{-2.571\ m}$$

$$\text{Power} = 0.0425\ \text{m}^3 \cdot \text{s}^{-1} \times 1000\ \text{kg} \cdot \text{m}^{-3} \times 9.81\ \text{N} \cdot \text{kg}^{-1} \times 2.772\ \text{m}$$

$$= \mathbf{1156\ W}$$

7.14 $$Re = \frac{ud}{\nu} = \frac{4Q}{\pi d \nu} = \frac{3.2}{\pi 0.6 \times 10^{-6} d}\ \text{(m)}$$

which is $>10^6$ for all reasonable value of d.

For given Q and ϱ, power $\propto h_f \propto \dfrac{f}{d^5}$

f depends on *relative* roughness; $\therefore f \propto \left(\dfrac{k}{d}\right)^{1/3}$

$$\therefore \text{Power} \propto d^{-16/3}$$

$\therefore$ If power halved, d multiplied by $2^{3/16} = 1.139$,

that is, $\boldsymbol{d = 569}$ **mm**

For n pipes in parallel $Re = \dfrac{4(Q/n)}{\pi d_2 \nu}$ $\quad \therefore d_2 = d_1/n$

$$\text{Total power} = n\varrho g \frac{Q}{n} h_f = \varrho g Q \frac{32 f l (Q/n)^2}{\pi^2 g d_2^5}$$

$$= \frac{32 \varrho Q^3 l}{\pi^2 d_2^5 n^2} \times \text{const} \left(\frac{k}{d_2}\right)^{1/3}$$

$$\propto \left(\frac{d_1}{n}\right)^{-16/3} n^{-2}$$

$\therefore \dfrac{\text{Power for } n \text{ pipes}}{\text{Power for 1 pipe}} = n^{10/3}$. This is >1 $\quad \therefore$ **No advantage.**

7.15 *Pipe* 1: $h_f + \dfrac{u_1^2}{2g}[\text{which is lost}] = \dfrac{u_1^2}{2g}\left(\dfrac{4 f l_1}{d_1} + 1\right)$

$$= \frac{16 Q_1^2}{\pi^2 d_1^4 2g}\left(\frac{4 f l_1}{d_1} + 1\right)$$

$$= \frac{16Q_1^2}{\pi^2(0.1)^4 19.62}\left(\frac{4 \times 0.008 \times 45}{0.1} + 1\right) \text{s}^2 \cdot \text{m}^{-5}$$

$$= 1.272 \times 10^4 Q_1^2 \text{ s}^2 \cdot \text{m}^{-5}$$

Pipe 2: $$h_f + \frac{u_2^2}{2g} = \frac{16Q_2^2}{\pi^2(0.15)^4 19.62}\left(\frac{4 \times 0.008 \times 60}{0.15} + 1\right) \text{s}^2 \cdot \text{m}^{-5}$$

$$= 2252 Q_2^2 \text{ s}^2 \cdot \text{m}^{-5}$$

∴ Total head at pump outlet (inlet as datum)

$$= 6.6 \text{ m} + 12\,720 Q_1^2 \text{ s}^2 \cdot \text{m}^{-5} = 8.6\text{m} + 2252 Q_2^2 \text{ s}^2 \cdot \text{m}^{-5}$$

$$\therefore 12\,720(0.037 - Q_2)^2 = 2 + 2252 Q_2^2$$

whence $Q_2 = 0.02153 \text{ m}^3 \cdot \text{s}^{-1}$

∴ Head at outlet $= \left\{8.6 + 2252(0.02153)^2\right\} \text{m} =$ **9.64 m**

7.16 $$h_{AD} = \frac{4 \times 0.01 \times 16 \times 10^3}{0.3 \times 19.62} u_A^2 = 108.7 u_A^2 \ [\text{s}^2 \cdot \text{m}^{-1}] = 125 \text{ m} - h$$

where h = head at D above C.

$$h_{DB} = \frac{4 \times 0.01 \times 9500}{0.2 \times 19.62} u_B^2 = 96.8 u_B^2 \ [\text{s}^2 \cdot \text{m}^{-1}] = h - 30 \text{ m}$$

$$h_{DC} = \frac{4 \times 0.01 \times 8000}{0.15 \times 19.62} u_C^2 = 108.7 u_C^2 \ [\text{s}^2 \cdot \text{m}^{-1}] = h$$

Continuity: $\frac{\pi}{4}(0.3)^2 u_A = \frac{\pi}{4}(0.2)^2 u_B + \frac{\pi}{4}(0.15)^2 u_C$

Hence $h = 89.2$ m.

$$\therefore u_B = \sqrt{\frac{59.2}{96.8}} \text{ m} \cdot \text{s}^{-1} = 0.781 \text{ m} \cdot \text{s}^{-1}; \quad u_C = \sqrt{\frac{89.2}{108.7}} \text{ m} \cdot \text{s}^{-1}$$

$$= 0.906 \text{ m} \cdot \text{s}^{-1}$$

$$\therefore Q_B = \frac{\pi}{4}(0.2)^2 0.781 \text{ m}^3 \cdot \text{s}^{-1} = \mathbf{0.02456 \text{ m}^3 \cdot \text{s}^{-1}}$$

$$Q_C = \frac{\pi}{4}(0.15)^2 0.906 \text{ m}^3 \cdot \text{s}^{-1} = \mathbf{0.0160 \text{ m}^3 \cdot \text{s}^{-1}}$$

7.17 Q from A is $0.48 \text{ m}^3 \cdot \text{s}^{-1}$ $\quad \therefore u_A = \frac{0.48}{(\pi/4)(0.75)^2} \text{ m} \cdot \text{s}^{-1}$

$$= 1.086 \text{ m} \cdot \text{s}^{-1}$$

$$(h_f)_A = \frac{4 \times 0.006 \times 10^4}{0.75}\frac{1.086^2}{19.62} \text{ m} = 19.25 \text{ m}$$

Head at junction above level of pipe

$$= \left\{3 + (2.2 \times 10^{-3})10^4 - 19.25\right\} \text{ m} = 5.75 \text{ m}$$

$$(h_f)_B = \left\{5.75 + \left(2.75 \times 10^{-3}\right) 5500 - 3\right\} \text{ m} = 17.875 \text{ m}$$

$$= \frac{4 \times 0.006 \times 5500}{d_B 19.62} \left(\frac{0.24}{(\pi/4)d_B^2}\right)^2 \text{ m}^6$$

whence $d_B =$ **0.512 m**

$$(h_f)_C = \left\{5.75 + \left(3.2 \times 10^{-3}\right) 3000 - 3\right\} \text{ m} = 12.35 \text{ m}$$

$$= \frac{4 \times 0.006 \times 3000}{d_C 19.62} \left(\frac{0.24}{(\pi/4)d_C^2}\right)^2 \text{ m}^6$$

whence $d_C =$ **0.488 m**

7.18 $30 \text{ m} = \dfrac{4 \times 0.008 \times 1000}{0.6} \dfrac{u^2}{19.62} \text{ s}^2 \cdot \text{m}^{-1} \quad \therefore u = 3.322 \text{ m} \cdot \text{s}^{-1}$

$$\therefore Q = \frac{\pi}{4}(0.6)^2 3.322 \text{ m}^3 \cdot \text{s}^{-1} = \mathbf{0.939\ m^3 \cdot s^{-1}}$$

$$\tau_0 = \frac{1}{2}\varrho u^2 f = \frac{1}{2} 1000 \times 3.322^2 \times 0.008 \text{ Pa} = \mathbf{44.1\ Pa}$$

In first half of pipe $u_x = u_{in} - \dfrac{u_{in} x}{1500}$

at x metres from inlet since $(-)\dfrac{dQ}{dx} = \dfrac{Q/3}{500 \text{ m}}$

$$h_{f1} = \frac{4f}{2gd}\int_0^{500\text{m}} u_x^2 dx = \frac{4f}{2gd}\int_0^{500\text{m}} \left(u_{in} - \frac{u_{in} x}{1500}\right)^2 dx$$

$$= \frac{0.032}{19.62 \times 0.6} u_{in}^2 \left[\frac{1}{3}\left(1 - \frac{x}{1500}\right)^3 (-1500)\right]_0^{500}$$

$$= 1.359 u_{in}^2 \left\{1^3 - \left(\frac{2}{3}\right)^3\right\} = 1.359 u_{in}^2 \frac{19}{27}$$

$$h_{f2} = \frac{4 \times 0.008 \times 500}{0.6 \times 19.62}\left(\frac{2}{3}u_{in}\right)^2 = 1.359 u_{in}^2 \frac{4}{9}$$

$$\therefore 30 \text{ m} = h_{f1} + h_{f2} = 1.359 u_{in}^2 \left(\frac{19}{27} + \frac{4}{9}\right)$$

whence $u_{in} = 4.385 \text{ m} \cdot \text{s}^{-1}$

$$\therefore u_{\text{out}} = \frac{2}{3} \times 4.385 \text{ m} \cdot \text{s}^{-1} = 2.923 \text{ m} \cdot \text{s}^{-1}$$

$$\therefore Q_{\text{out}} = 2.923\frac{\pi}{4}(0.6)^2 \text{ m}^3 \cdot \text{s}^{-1} = \mathbf{0.826\ m^3 \cdot s^{-1}}$$

7.19

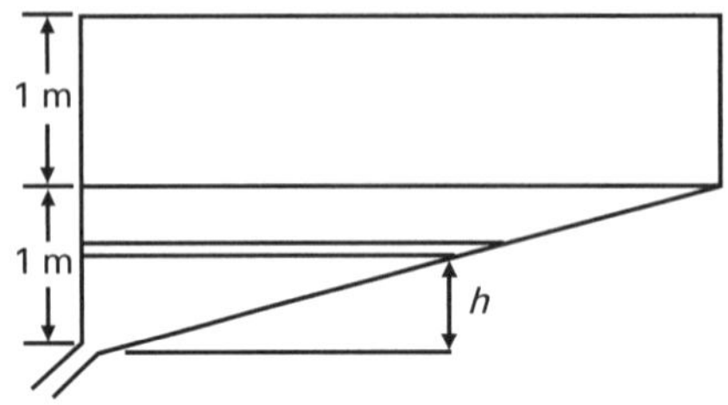

For upper, rectangular, portion $-(18 \times 9 \text{ m}^2)\,dh = 0.9\frac{\pi}{4}(0.15 \text{ m})^2 \times 2\sqrt{2gh}\,dt$

$$\therefore t_1 = \frac{-18 \times 9}{0.9\,(\pi/4) \times 0.15^2 \times 2\sqrt{19.62}} \int_{2\text{m}}^{1\text{m}} h^{-1/2}\,dh = 952 \text{ s}$$

For lower, triangular, portion $A = \frac{h}{1} 18 \times 9 \text{ m}^2 = 162h \text{ m}^2$

$$t_2 = \frac{1}{0.9(\pi/4) \times 0.15^2 \times 2\sqrt{19.62}} \int_{1\text{m}}^{0} -Ah^{-1/2}\,dh$$

$$= 7.10 \int_{1\text{m}}^{0} -162h^{1/2}dh = 767 \text{ s}$$

Total time $= t_1 + t_2 = \mathbf{1719\ s}$

7.20 $\delta h = \delta\,(z_A - z_B) = \delta z_A - \left(-\frac{1.5}{7.5}\delta z_A\right) = \frac{6}{5}\delta z_A \quad \therefore \delta z_A = \frac{5}{6}\delta h$

$$-(1.5 \text{ m}^2)\frac{5}{6}\delta h = 0.6\frac{\pi}{4}(0.025 \text{ m})^2\sqrt{19.62 \text{ m} \cdot \text{s}^{-2}}\,h^{1/2}\delta t$$

$$\therefore t = \frac{-1.5 \times (5/6)}{0.6(\pi/4) \times 0.025^2\sqrt{19.62}} \int_{1.2\text{m}}^{0} h^{-1/2}\,dh = 958 \times 2\sqrt{1.2} \text{ s}$$

$$= \mathbf{2100\ s}$$

7.21 Let surface areas of tanks be A, B. Then $A(-\delta z_A) = B\delta z_B = Q\delta t$

$$\delta h = \delta\,(z_A - z_B) = \delta z_A\left(1 + \frac{A}{B}\right) = \delta z_A\left(1 + \frac{2.5^2}{1.5^2}\right) = \frac{34}{9}\delta z_A$$

$$\therefore Q\delta t = -\frac{9}{34}A\delta h = -(1.30 \text{ m}^2)\delta h$$

$$h = \left(\frac{4fl}{d} + 1.5\right)\frac{u^2}{2g} = \left(\frac{4 \times 0.01 \times 75}{0.05} + 1.5\right)\frac{u^2}{2g} = 61.5\frac{u^2}{2g}$$

$$\therefore u = \left(0.565\ \mathrm{m^{1/2} \cdot s^{-1}}\right) h^{1/2} \quad \text{and} \quad Q = \left(0.00111\ \mathrm{m^{5/2} \cdot s^{-1}}\right) h^{1/2}$$

$$t = -\int_{1\mathrm{m}}^{x} \frac{1.30}{0.00111 h^{1/2}}\, \mathrm{d}h = \frac{2.60}{0.00111}\left(1 - x^{1/2}\right) = 1200\ \mathrm{s}$$

whence $x = 0.238$ m

$$\therefore \Delta h = 0.762\ \mathrm{m} \qquad \therefore \Delta z_{\mathrm{A}} = \frac{9}{34} \times 0.762\ \mathrm{m} = \mathbf{0.2017\ m}$$

7.22

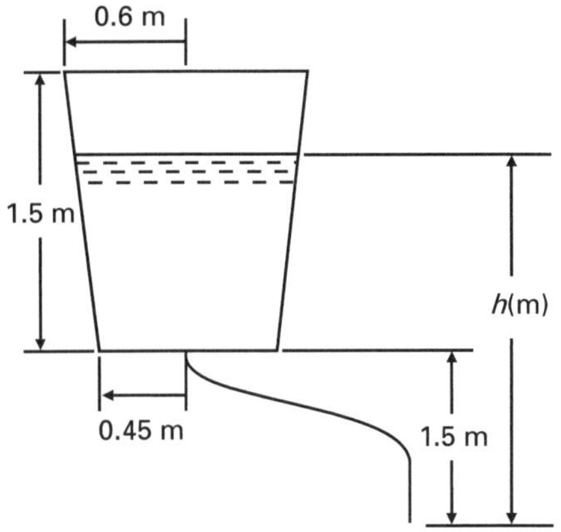

$$A = \pi r^2$$
$$= \pi\left\{0.45 + (h - 1.5) \times \frac{0.15}{1.5}\right\}^2\ \mathrm{m^2}$$
$$= \pi\,(0.3 + 0.1h)^2\ \mathrm{m^2}$$

$$t = -\int_{h_1}^{h_2} \frac{A}{au}\mathrm{d}h = \int_{h_1}^{h_2} \frac{(0.3 + 0.1h)^2}{d^2/4}\sqrt{\frac{4fl}{2gd}}\, h^{-1/2}\mathrm{d}h$$

$$= \frac{4}{d^2}\sqrt{\frac{4fl}{2gd}}\int_{1.7}^{2.8}(0.09h^{-1/2} + 0.06h^{1/2} + 0.01h^{3/2})\ \mathrm{d}h$$

$$= \frac{4}{d^2}\sqrt{\frac{4fl}{2gd}}\left[0.18h^{1/2} + 0.04h^{3/2} + 0.004h^{5/2}\right]_{1.7}^{2.8}$$

$$= \frac{4}{d^2}\sqrt{\frac{4fl}{2gd}} \times 0.2027 = 600\ \mathrm{s}$$

$$\therefore d = \left\{\frac{4^2 \times 4 \times 0.008 \times 36 \times 0.2027^2}{19.62 \times 600^2}\right\}^{1/5}\ \mathrm{m}$$

$= \mathbf{0.0404\ m, say\ 40\ mm}$

7.23

$$\text{Initially } u = \frac{0.017\ \mathrm{m^3 \cdot s^{-1}}}{(\pi/4)(0.075\ \mathrm{m})^2} = 3.848\ \mathrm{m \cdot s^{-1}}$$

$$1.5\ \mathrm{m} + h_1 = \frac{u^2}{2g}\left(1.0 + \frac{4 \times 0.01 \times 3}{0.075}\right) = 2.6\frac{u^2}{2g}$$

whence $h_1 = 0.462$ m (depth in tank)

$$A = \frac{\pi}{4}\left(4.25 + \frac{1.75}{1.5}h\right)^2 \ \text{m}^2$$

When $Q = 0.034\ \text{m}^3 \cdot \text{s}^{-1}$, $\left(0.034\ \text{m}^3 \cdot \text{s}^{-1}\right) \text{d}t = A\text{d}h + au\ \text{d}t$

that is,
$$\text{d}t\left[0.034 - \frac{\pi}{4}(0.075)^2\left\{\frac{19.62(1.5+h)}{2.6}\right\}^{1/2}\right]$$

$$= \frac{\pi}{4}\left(4.25 + \frac{7}{6}h\right)^2 \text{d}h\,[\text{metre, second units}]$$

$$\therefore t = \frac{\pi}{4}\int_{0.462}^{1.5} \frac{(4.25 + (7/6)h)^2 \text{d}h}{0.034 - 0.01214(1.5+h)^{1/2}} \ \text{seconds}$$

An analytical solution is possible. Put denominator $= x$,

then $\left(\dfrac{0.034 - x}{0.01214}\right) = 1.5 + h$

and $\text{d}h = \dfrac{-2(0.034 - x)}{0.01214^2}\,\text{d}x$

$$\text{Then } t = -\frac{\pi}{4}\frac{2}{0.01214^2}\int_{0.017}^{0.01298}\left[4.25 + \frac{7}{6}\left\{\left(\frac{0.034 - x}{0.01214}\right)^2 - 1.5\right\}\right]^2$$

$$\times\,(0.034 - x)\,\frac{\text{d}x}{x}$$

$$= 1.067 \times 10^4 \int_{0.01298}^{0.017}\left(11.66 - 539x + 7921x^2\right)^2$$

$$\times\left(\frac{0.034}{x} - 1\right)\text{d}x$$

$$= 1.067 \times 10^4 \int_{0.01298}^{0.017}\left(\frac{4.62}{x} - 563 + 2.870 \times 10^4 x\right.$$

$$\left. - \,7.65 \times 10^5 x^2 + 1.067 \times 10^7 x^3 - 6.27 \times 10^7 x^4\right)\text{d}x$$

$$= 1.067 \times 10^4 \Big[4.62 \ln x - 563x + 1.435 \times 10^4 x^2$$

$$- \,2.55 \times 10^5 x^3 + 2.667 \times 10^6 x^4 - 1.255 \times 10^7 x^5\Big]_{0.01298}^{0.017}$$

$= \mathbf{1626\ s}$ (This depends on relatively small differences.)

7.24

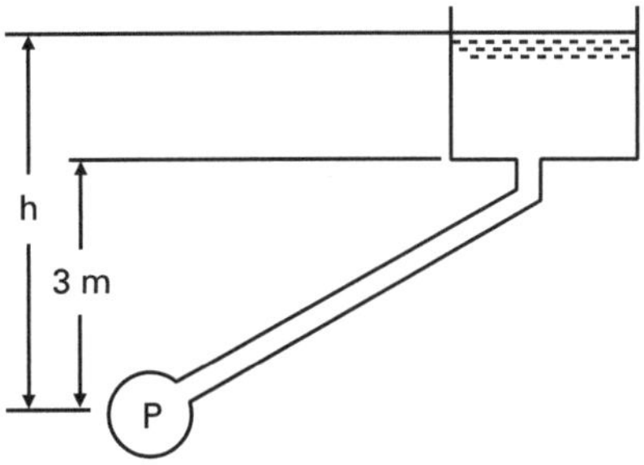

At any time, head required

$$= h + h_f = h + \frac{32 f l Q^2}{\pi^2 g d^5}$$

$$= \frac{500 \times 10^3}{1000 \times 9.81}\ \text{m} = 50.97\ \text{m}$$

$$\therefore Q^2 = (50.97\ \text{m} - h)\,\frac{\pi^2 g d^5}{32 f l}$$

$$= (50.97\ \text{m} - h)\,2.955 \times 10^{-6}\ \text{m}^5 \cdot \text{s}^{-2}$$

Also $Q = A\mathrm{d}h/\mathrm{d}t$

$$\therefore t = \int_{3.2\text{m}}^{5.5\text{m}} \frac{5\mathrm{d}h 10^3}{\sqrt{2.955}\sqrt{50.97 - h}} = -\frac{5000}{\sqrt{2.955}}\left[2\,(50.97 - h)^{1/2}\right]_{3.2}^{5.5}\ \text{s}$$

$$= \frac{10^4}{\sqrt{2.955}}\left(\sqrt{47.77} - \sqrt{45.47}\right)\text{s} = \mathbf{980\ s}$$

$$\text{Power required} = \frac{1}{\eta} Q \varrho g (h + h_f) = \frac{1}{\eta} Q (500\ \text{kPa})$$

$$\therefore \text{Electricity used in time } \delta t \text{ is } \frac{1}{\eta} Q (500\ \text{kPa}) \delta t$$

$$\therefore \text{Total used} = \frac{1}{\eta}(500\ \text{kPa}) \int Q \mathrm{d}t$$

$$= \frac{1}{\eta}(500\ \text{kPa})(\text{Volume transferred})$$

$$= \frac{1}{0.52} 500 \times 10^3 \times 5 \times 2.3\ \text{N} \cdot \text{m} = \frac{5.75 \times 10^6}{0.52 \times 10^3}\ \text{kW} \cdot \text{s}$$

$$= \frac{5.75 \times 10^3}{0.52 \times 3600}\ \text{kW} \cdot \text{h} = \mathbf{3.072\ kW \cdot h}$$

7.25 At any instant, head required $= h + h_f = h + \dfrac{32 f l Q^2}{\pi^2 g d^5}$

$$= \frac{300 \times 10^3}{1000 \times 9.81}\ \text{m} = 30.58\ \text{m}$$

$$\therefore Q^2 = (30.58\ \text{m} - h)\,\frac{\pi^2 g d^5}{32 f l}$$

Also $Q = A\mathrm{d}h/\mathrm{d}t$

$$\therefore t = \int_{3.0\mathrm{m}}^{5.6\mathrm{m}} \frac{A\mathrm{d}h}{\pi(30.58 - h)^{1/2}}\sqrt{\frac{32fl}{gd^5}}$$

$$= -\frac{A}{\pi}\sqrt{\frac{32fl}{gd^5}}\left[2(30.58 - h)^{1/2}\right]_{3.0}^{5.6}$$

$$= \frac{2A}{\pi}\sqrt{\frac{32fl}{gd^5}}(27.58^{1/2} - 24.98^{1/2})$$

$$\therefore d^{5/2} = \frac{2A}{\pi t}\sqrt{\frac{32fl}{g}}0.2537 \not< \frac{12}{15 \times 60\pi}\sqrt{\frac{32 \times 0.007 \times 35}{9.81}}0.2537$$

$$= 9.624 \times 10^{-4}\left[\mathrm{m}^{5/2}\right]$$

$d \not< 0.0621$ m that is, $\quad$ **$d > 62.1$ mm**

7.26 Final difference of levels = $(40 - 2 \times 15)$ mm = 10 mm

$$A\left(-\frac{\mathrm{d}z_1}{\mathrm{d}t}\right) = Q = A\frac{\mathrm{d}z_2}{\mathrm{d}t}$$

$$\therefore \frac{\mathrm{d}h}{\mathrm{d}t} = \frac{\mathrm{d}}{\mathrm{d}t}(z_1 - z_2) = 2\frac{\mathrm{d}z_1}{\mathrm{d}t} = -2\frac{Q}{A} = -\frac{2g\pi d^4 h}{A128\nu l}$$

$$\therefore t = -\frac{64\nu lA}{g\pi d^4}\int_{0.04\mathrm{m}}^{0.01\mathrm{m}} \frac{\mathrm{d}h}{h} = \frac{64\nu lA}{g\pi d^4}\ln 4$$

$$\therefore \mu = \frac{\varrho g\pi d^4 t}{64lA\ln 4} = \frac{840 \times 9.81 \times 10^{-12} \times 478}{64 \times 0.4 \times \frac{1}{4}(0.02)^2\ \ln 4}\ \mathrm{Pa\cdot s}$$

$$= \mathbf{1.110\ mPa\cdot s}$$

7.27 $\Delta p^* = (7 \times 10^5 - 4.62 \times 10^5)$ Pa $- 1000$ kg·m^{-3}

$\times\ 9.81$ N·kg^{-1} $(30 \sin 45°)$ m = 29 900 Pa

$$\tau_0 = \frac{R}{2}\frac{\mathrm{d}p^*}{\mathrm{d}x} = 0.0125\ \mathrm{m}\left(-\frac{29\,900\ \mathrm{Pa}}{30\ \mathrm{m}}\right)$$

= (−)**12.46 Pa**

$$\tau = \frac{r}{2}\frac{\mathrm{d}p^*}{\mathrm{d}x} = 0.005\ \mathrm{m}\left(-\frac{29\,900\ \mathrm{Pa}}{30\ \mathrm{m}}\right)$$

= (−)**4.98 Pa**

7.28 $(p_1 - p_2)\pi R^2 - F =$ Rate of increase of momentum

$$= \int_0^R \varrho 2\pi r \, \mathrm{d}r u^2 - \varrho\pi R^2 V^2$$

$$= 2\pi\varrho 4V^2 \int_0^R r\left(1 - \frac{r^2}{R^2}\right)^2 \mathrm{d}r - \varrho\pi R^2 V^2$$

$$= \frac{4}{3}\pi\varrho R^2 V^2 - \varrho\pi R^2 V^2 = \frac{1}{3}\pi R^2 \varrho V^2$$

$$\therefore \boldsymbol{F = \pi R^2(p_1 - p_2 - \varrho V^2/3)}$$

Chapter 8

8.1 $$\frac{\delta^*}{\delta} = \int_0^1 \left(1 - \frac{u}{u_m}\right) d\eta - \int_0^1 \left(1 - \sin\frac{\pi\eta}{2}\right) d\eta - \left[\eta + \frac{2}{\pi}\cos\frac{\pi\eta}{2}\right]_0^1$$

$$= 1 - \frac{2}{\pi} = \mathbf{0.3634}$$

$$\frac{\theta}{\delta} = \int_0^1 \frac{u}{u_m}\left(1 - \frac{u}{u_m}\right) d\eta = \int_0^1 \left(\sin\frac{\pi\eta}{2} - \sin^2\frac{\pi\eta}{2}\right) d\eta$$

$$= \left[-\frac{2}{\pi}\cos\frac{\pi\eta}{2} - \frac{\eta}{2} + \frac{1}{2\pi}\sin\pi\eta\right]_0^1 = -\frac{1}{2} + \frac{2}{\pi} = \mathbf{0.1366}$$

8.2 From Blasius's solution, $\delta \simeq 4.91x/(Re_x)^{1/2}$

$$Re_x = \frac{6\ \mathrm{m\cdot s^{-1}} \times 2.4\ \mathrm{m}}{14.9 \times 10^{-6}\ \mathrm{m^2\cdot s^{-1}}} = 9.66 \times 10^5$$

$$\therefore \delta \simeq \frac{4.91 \times 2.4\ \mathrm{m}}{(9.66 \times 10^5)^{1/2}}$$

$$= \mathbf{0.01199\ m}$$

$$\tau_0 = 0.332\ \varrho u_m^2 (Re_{x/2})^{-1/2} \quad \text{[Table 8.1]}$$

$$= 0.332 \times 1.21\ \mathrm{kg\cdot m^{-3}} \times \left(6\ \mathrm{m\cdot s^{-1}}\right)^2 \left(\frac{9.66 \times 10^5}{2}\right)^{-1/2}$$

$$= \mathbf{0.0208\ Pa}$$

Total drag on one side $= 0.9\ \mathrm{m} \times 0.664 \times 1.21\ \mathrm{kg\cdot m^{-3}} \times (6\ \mathrm{m\cdot s^{-1}})^2$

$$\times \frac{2.4\ \mathrm{m}}{(9.66 \times 10^5)^{1/2}} = 0.0636\ \mathrm{N}$$

$$\therefore \text{Power} = 2 \times 0.0636\ \mathrm{N} \times 6\ \mathrm{m\cdot s^{-1}} = \mathbf{0.763\ W}$$

For turbulent layer

$$C_F = 0.074(Re_l)^{-1/5} = \frac{0.074}{(9.66)^{1/5} \times 10} = 0.0470$$

$$\therefore \text{ Power} = 2 \times \frac{1}{2}\varrho u_{\mathrm{m}}^2(\text{Area})C_{\mathrm{F}}u_{\mathrm{m}}$$
$$= 1.21 \text{ kg}\cdot\text{m}^{-3}(6 \text{ m}\cdot\text{s}^{-1})^3(2.4 \text{ m} \times 0.9 \text{ m})0.00470$$
$$= \mathbf{2.654\ W}$$

8.3 Since $U =$ constant, eqn 8.9 gives

$$\tau_0 = \varrho\frac{\partial}{\partial x}\int_0^{\delta}(U-u)u\,\mathrm{d}y = \varrho U^2\frac{\partial}{\partial x}\delta\int_0^1 \eta^{1/7}(1-\eta^{1/7})\mathrm{d}\eta$$

[where $\eta = y/\delta$]

$$= \varrho U^2\frac{\partial}{\partial x}\delta\left[\frac{7}{8}\eta^{8/7} - \frac{7}{9}\eta^{9/7}\right]_0^1 = \frac{7}{72}\varrho U^2\frac{\mathrm{d}\delta}{\mathrm{d}x}$$
$$= 0.023\varrho U^2\left(\frac{\nu}{U\delta}\right)^{1/4}$$
$$\therefore \frac{7}{72}\delta^{1/4}\mathrm{d}\delta = 0.023\left(\frac{\nu}{U}\right)^{1/4}\mathrm{d}x$$
$$\therefore \frac{7}{72}\times\frac{4}{5}\delta^{5/4} + \text{const} = 0.023\left(\frac{\nu}{U}\right)^{1/4}x$$

but const. $= 0$ since $\delta = 0$ when $x = 0$

$$\therefore \delta = 0.3773x^{4/5}(\nu/U)^{1/5}$$

$$\text{Total drag } = 2b\int_0^x \tau_0\,\mathrm{d}x = 2b \times \frac{7}{72}\varrho U^2\delta$$
$$= \frac{7}{36}b\varrho U^2 0.3773x^{4/5}(\nu/U)^{1/5}$$
$$= \frac{7}{36}0.8 \text{ m}(1000 \text{ kg}\cdot\text{m}^{-3})(3 \text{ m}\cdot\text{s}^{-1})^2$$
$$\times 0.3773(2.5 \text{ m})^{4/5}\left(\frac{10^{-6} \text{ m}^2\cdot\text{s}^{-1}}{3 \text{ m}\cdot\text{s}^{-1}}\right)^{1/5}$$
$$= \mathbf{55.69\ N}$$

Power $= 55.69 \text{ N} \times 3 \text{ m}\cdot\text{s}^{-1} =$ **167.1 W**

8.4 (a) $Re_l = \dfrac{10.5 \text{ m}\cdot\text{s}^{-1} \times 3 \text{ m}}{15 \times 10^{-6} \text{ m}^2\cdot\text{s}^{-1}} = 2.1 \times 10^6 \quad Re_{\mathrm{t}} = 5 \times 10^5$

$$\therefore x_{\mathrm{t}}/l = 5/21$$

$$\frac{x_{\mathrm{t}} - x_0}{x_{\mathrm{t}}} = \frac{36.9}{(5 \times 10^5)^{3/8}} = 0.269 \quad \text{[eqn 8.29]}$$

$$\therefore x_0/x_{\mathrm{t}} = 0.731$$

$$\therefore \quad \frac{x_0}{l} = \frac{x_0}{x_t}\frac{x_t}{l} = 0.731 \times \frac{5}{21} = 0.1740$$

$$C_F = \frac{0.074}{(2.1 \times 10^6)^{1/5}}[1 - 0.174]^{4/5} = 0.003454$$

$$\therefore \text{ Drag} = \frac{1}{2}\varrho u_m^2(\text{Area})0.003454$$

(b) $$Re_l = \frac{10.5 \times 0.3}{15 \times 10^{-6}} = 2.1 \times 10^5 \quad \therefore \text{Wholly laminar}$$

$$C_F = \frac{2\theta}{x} = 1.328(Re_l)^{-1/2} \quad [\text{Table 8.1}]$$

$$= 0.00290$$

$$\therefore \text{ Ratio} = \frac{0.003454}{0.00290} = \mathbf{1.192}$$

8.5 $$u_m = \frac{160 \times 10^3}{3600} \text{ m} \cdot \text{s}^{-1} = 44.44 \text{ m} \cdot \text{s}^{-1}$$

$$\therefore \; Re_l = \frac{44.44 \times 110 \times 1.22}{1.79 \times 10^{-5}} = 3.332 \times 10^8$$

If transition occurs at $Re = 5 \times 10^5$, length of laminar layer

$$= \frac{5 \times 10^5}{3.332 \times 10^8} \times 110 \text{ m} = \mathbf{0.165 \text{ m}}$$

$\therefore$ Assume layer turbulent throughout.

From eqn 8.26, $$C_F = \frac{0.455}{(8.5227)^{2.58}} = 0.001808$$

$$\text{Power} = 110 \text{ m} \times 8.25 \text{ m} \times \frac{1}{2} \times 1.22 \text{ kg} \cdot \text{m}^{-3}$$

$$\times (44.44 \text{ m} \cdot \text{s}^{-1})^2 0.001808 \times 44.44 \text{ m} \cdot \text{s}^{-1}$$

$$= \mathbf{87.9 \text{ kW}}$$

8.6

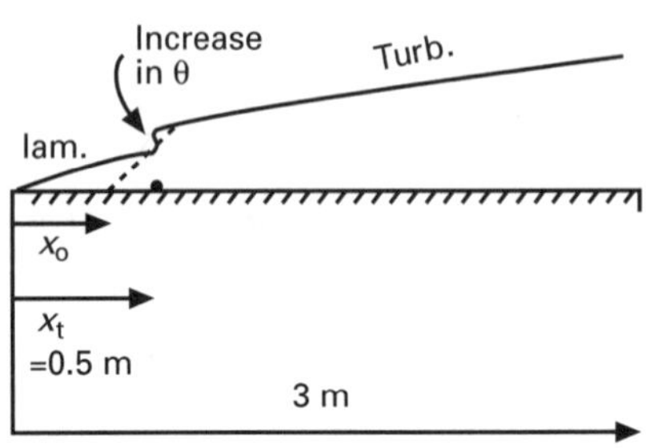

At transition Re_{x_t}

$$= \frac{29 \text{ m} \cdot \text{s}^{-1} \times 0.5 \text{ m}}{14.5 \times 10^{-6} \text{ m}^2 \cdot \text{s}^{-1}}$$

$$= 10^6$$

$$(\theta_t)_{lam} = \frac{0.664 \times 0.5 \text{ m}}{(10^6)^{1/2}}$$

$$= 0.332 \text{ mm}$$

So for turbulent layer with virtual origin at $x = x_0$

$$1.15 \times 0.332 \times 10^{-3}\ \text{m} = \frac{0.037(x_t - x_0)}{\{U(x_t - x_0)/\nu\}^{1/5}}$$

$$= \frac{0.037(x_t - x_0)^{4/5} x_t^{1/5}}{(Ux_t/\nu)^{1/5}}$$

whence $$x_t - x_0 = \left\{\frac{10^{6/5} \times 1.15 \times 0.332 \times 10^{-3}\ \text{m}}{0.037(0.5\ \text{m})^{1/5}}\right\}^{5/4}$$

$$= 0.1237\ \text{m}$$

$\therefore\ x_0 = 0.3763\ \text{m}$

At $x = 3$ m $$\theta = \frac{0.037(3 - 0.3763)\ \text{m}}{\{29(3 - 0.3763)/14.5 \times 10^{-6}\}^{1/5}} = 0.004397\ \text{m}$$

$$C_D = \frac{\text{Drag/width}}{\frac{1}{2}\varrho U^2 \times \text{length}} = \frac{\varrho U^2 \theta_{\text{trailing edge}}}{\frac{1}{2}\varrho U^2 \times \text{length}} = \frac{2\theta_{\text{t.e.}}}{\text{length}}$$

$$= \frac{2 \times 0.004397\ \text{m}}{3\ \text{m}} = \mathbf{0.00293}$$

8.7 If main-stream velocity remained unchanged,

$$Re_l = \frac{1.8 \times 0.150}{1.21 \times 10^{-6}} = 2.231 \times 10^5$$

$\therefore$ Laminar boundary layer

$$\delta^* = \frac{1.721 \times 0.150\ \text{m}}{(2.231 \times 10^5)^{1/2}}\quad \text{[Table8.1]} = \mathbf{0.546\ mm}$$

$$\text{Outlet velocity} = u_1 \frac{b^2}{(b - 2\delta^*)^2} = 1.8\ \text{m}\cdot\text{s}^{-1}\frac{25^2}{(25 - 1.092)^2}$$

$$= \mathbf{1.968\ m \cdot s^{-1}}$$

$$\text{Pressure drop} = \frac{1}{2}1000\ \text{kg}\cdot\text{m}^{-3}(1.968^2 - 1.8^2)\ \text{m}^2\cdot\text{s}^{-2}$$

$$= \mathbf{317.2\ Pa}$$

8.8 $$Re = \frac{22 \times 0.003}{15 \times 10^{-6}} = 4400$$

Eqn 8.33 : $$\frac{\omega \times 0.003\ \text{m}}{22\ \text{m}\cdot\text{s}^{-1}} = 0.198\left(1 - \frac{19.7}{4400}\right) \quad \therefore\ \omega = \mathbf{1445\ Hz}$$

8.9 $p = \text{const} - \frac{1}{2}\varrho(2U\sin\theta)^2$ if boundary layer thin.
If separation occurs at $\theta = \phi$, total component of pressure drag where boundary layer is attached

$$= 2\int_0^{\phi} (\text{const} - 2\varrho U^2 \sin^2\theta) r d\theta \cos\theta \text{ per unit length}$$

$$= 2r\left[C\sin\theta - 2\varrho U^2 \frac{1}{3}\sin^3\theta\right]_0^{\phi}$$

$$= 2r\left[C\sin\phi - \frac{2}{3}\varrho U^2 \sin^3\phi\right]$$

p at separation $= C - 2\varrho U^2 \sin^2\phi$
∴ Drag divided by length, due to pressure in wake

$$= -2r(C - 2\varrho U^2 \sin^2\phi)\sin\phi$$

∴ Net pressure drag/length

$$= 4r\varrho U^2 \sin^3\phi - \frac{4}{3}r\varrho U^2 \sin^3\phi = \frac{8}{3}r\varrho U^2 \sin^3\phi$$

$$C_D = \frac{\frac{8}{3}r\varrho U^2 \sin^3\phi}{\frac{1}{2}\varrho U^2 \times 2r} = \frac{8}{3}\sin^3\phi = 1.24 \quad \therefore \ \sin\phi = 0.7747$$

$\therefore \ \phi = 50.78°$ or $129.22°$

But separation occurs only where pressure gradient adverse.

$\therefore \ \phi = 129.2°$, **skin friction neglected.**

8.10 $u_m = 2U\sin\frac{x}{r}$

$$\therefore \ \frac{du_m}{dx} = \frac{2}{r}U\cos\frac{x}{r}$$

At $x = x_1$,
$$\theta^2 = \frac{0.45\nu}{2U\sin^6(x_1/r)}\int_0^{x_1}\sin^5\frac{x}{r}\,dx$$

$$= \frac{0.45\nu r}{2U\sin^6(x_1/r)}\int_0^{x_1} -\left(1-\cos^2\frac{x}{r}\right)^2 d\left(\cos\frac{x}{r}\right)$$

$$= \frac{-0.45\nu r}{2U\sin^6(x_1/r)}\left[\cos\frac{x}{r} - \frac{2}{3}\cos^3\frac{x}{r} + \frac{1}{5}\cos^5\frac{x}{r}\right]_0^{x_1}$$

$$= \frac{0.45\nu r}{2U\sin^6(x_1/r)} \times \left[\frac{8}{15} - \cos\frac{x_1}{r} + \frac{2}{3}\cos^3\frac{x_1}{r} - \frac{1}{5}\cos^5\frac{x_1}{r}\right]$$

$$\therefore \lambda = \frac{\theta^2}{\nu}\frac{du_m}{dx}$$

$$= \frac{0.45r}{2U\sin^6(x_1/r)}\left[\frac{8}{15} - \cos\frac{x_1}{r} + \frac{2}{3}\cos^3\frac{x_1}{r} - \frac{1}{5}\cos^5\frac{x_1}{r}\right]$$

$$\times \frac{2}{r}U\cos\frac{x_1}{r}$$

which is satisfied by $\lambda = -0.09$ and $\frac{x_1}{r} \equiv$ **103.1°**

8.11 $Re = \dfrac{60 \times 0.150}{15 \times 10^{-6}} = 6 \times 10^5$

(a) From Fig. 8.14, $C_D \simeq 0.08$

$$\therefore \text{Drag} \simeq \frac{1}{2} \times 1.21 \text{ kg}\cdot\text{m}^{-3}(60 \text{ m}\cdot\text{s}^{-1})^2\frac{\pi}{4}(0.15 \text{ m})^2 0.08$$

$$= \mathbf{3.08\ N}$$

(b) From Fig. 8.14, $C_D \simeq 1.10$

$$\therefore \text{Drag} \simeq \frac{1}{2} \times 1.21 \times 60^2\frac{\pi}{4}0.15^2 \times 1.10 \text{ N} = \mathbf{42.0\ N}$$

8.12 Drag $= 90 \times 9.81$ N when $u = 6 \text{ m}\cdot\text{s}^{-1}$
Assume $C_D = 1.32$. Then

$$90 \times 9.81 \text{ N} = \frac{1}{2} \times 1.22 \text{ kg}\cdot\text{m}^{-3}(6 \text{ m}\cdot\text{s}^{-1})^2 \times 1.32 \times \frac{\pi}{4}d^2$$

$\therefore d =$ **6.23 m**

Check: $Re = \dfrac{6 \times 6.23}{15 \times 10^{-6}}$ exceeds 10^3

8.13 $$(4Re/3C_D)^{1/3} = u_T\left\{\varrho^2/g\mu(\Delta\varrho)\right\}^{1/3}$$

$$= 10 \text{ m}\cdot\text{s}^{-1}\left\{\left(1.21 \text{ kg}\cdot\text{m}^{-3}\right)^2 \Big/ \left(9.81 \text{ N}\cdot\text{kg}^{-1}\right.\right.$$

$$\left.\left.\times 18.0 \times 10^{-6} \text{ N}\cdot\text{s}\cdot\text{m}^{-2} \times 2798.79 \text{ kg}\cdot\text{m}^{-3}\right)\right\}^{1/3}$$

$$= 14.36$$

$\therefore$ From Fig. 8.15, $Re = 1148$

$$\therefore d = \frac{1148 \times 18 \times 10^{-6} \text{ N}\cdot\text{s}\cdot\text{m}^{-2}}{10 \text{ m}\cdot\text{s}^{-1} \times 1.21 \text{ kg}\cdot\text{m}^{-3}} = \mathbf{1.708\ mm}$$

For water $\left\{\frac{3}{4}C_D(Re)^2\right\}^{1/3} = d\left\{\varrho(\Delta\varrho)g/\mu^2\right\}^{1/3}$

$$= 0.001708\ \text{m}\left\{\frac{1000\ \text{kg}\cdot\text{m}^{-3}\times 1800\ \text{kg}\cdot\text{m}^{-3}\times 9.81\ \text{N}\cdot\text{kg}^{-1}}{(10^{-3}\ \text{N}\cdot\text{s}\cdot\text{m}^{-2})^2}\right\}^{1/3}$$

$$= 44.48$$

$\therefore$ From Fig. 8.15, $Re = 443$

$$\therefore\ u_T = \frac{443\times 10^{-3}\ \text{N}\cdot\text{s}\cdot\text{m}^{-2}}{0.001708\ \text{m}\times 1000\ \text{kg}\cdot\text{m}^{-3}} = \mathbf{0.259\ m\cdot s^{-1}}$$

8.14 $$\overline{u} = \frac{0.0185\ \text{m}^3\cdot\text{s}^{-1}}{(\pi/4)(0.1\ \text{m})^2} = 2.355\ \text{m}\cdot\text{s}^{-1}$$

$$f = \frac{19.62\ \text{m}\cdot\text{s}^{-2}\times 1.89\ \text{m}\times 0.1\ \text{m}}{4\times 25\ \text{m}\times(2.355\ \text{m}\cdot\text{s}^{-1})^2} = 0.00668$$

$$Re = \frac{2.355\ \text{m}\cdot\text{s}^{-1}\times 0.1\ \text{m}}{1.2\times 10^{-6}\ \text{m}^2\cdot\text{s}^{-1}} = 1.963\times 10^5$$

Eqn 8.56:

$$(0.00668)^{-1/2} = -4\log\left\{\frac{k}{3.71d} + \frac{1.26}{1.963\times 10^5(0.00668)^{1/2}}\right\}$$

whence $k/d = \mathbf{0.002955}$

$$(u_m - \overline{u})\pi R^2 = \int_0^R (u_m - u)2\pi r\,dr$$

$$= -2\pi A\left(\frac{\tau_0}{\varrho}\right)^{1/2}\int_R^0\left(\ln\frac{R}{y}\right)(R-y)dy \quad \text{[from eqn 8.50]}$$

$$= 2\pi\times 2.5\overline{u}\left(\frac{f}{2}\right)^{1/2}\left(\frac{3R^2}{4}\right)$$

$$\therefore\ u_m - \overline{u} = 3.75\overline{u}(f/2)^{1/2}$$

$$\therefore\ u_m = \overline{u}\left\{1 + 3.75(f/2)^{1/2}\right\} = \mathbf{2.866\ m\cdot s^{-1}}$$

$$\tau_0 = \frac{1}{2}\varrho\overline{u}^2 f = \frac{1}{2}\times 1000\ \text{kg}\cdot\text{m}^{-3}(2.355\ \text{m}\cdot\text{s}^{-1})^2 0.00668$$

$$= \mathbf{18.54\ Pa}$$

8.15 $$\frac{h_f}{l} = \frac{1000\ \text{kg}\cdot\text{m}^{-3}}{0.7\ \text{kg}\cdot\text{m}^{-3}}\times\frac{50\times 10^{-3}\ \text{m}}{1000\ \text{m}} = 0.07143$$

$$Q = \frac{\pi}{4}d^2\overline{u} = \frac{\pi}{4}d^2\left(\frac{2gh_f d}{4fl}\right)^{1/2}$$

$$= \frac{\pi}{4}d^{5/2}\left(\frac{2gh_f}{4l}\right)^{1/2}\left\{4\log_{10}(Re\,f^{1/2}) - 0.4\right\}$$

$$= \frac{\pi}{4} d^{5/2} \left(\frac{gh_\mathrm{f}}{2l}\right)^{1/2} \left\{2\log_{10}(Re^2 f) - 0.4\right\}$$

$$= \frac{\pi}{4} d^{5/2} \left(\frac{gh_\mathrm{f}}{2l}\right)^{1/2} \left\{2\log_{10}\left(\frac{2gd^3h_\mathrm{f}}{4\nu^2 l}\right) - 0.4\right\}$$

$$= \frac{\pi}{2}(0.25\ \mathrm{m})^{5/2} \left(\frac{9.81\ \mathrm{m\cdot s^{-2}} 0.07143}{2}\right)^{1/2}$$

$$\times \left\{\log_{10}\left(\frac{9.81 \times 0.25^3 \times 0.07143}{2 \times 18^2 \times 10^{-12}}\right) - 0.2\right\}$$

$$= \mathbf{0.204\ m^3 \cdot s^{-1}}$$

$$\tau_0 = \frac{\Delta p^*}{l}\frac{d}{4} = \frac{\varrho g h_\mathrm{f}}{l}\frac{d}{4} = 0.7\ \mathrm{kg\cdot m^{-3}} \times 9.81\ \mathrm{N\cdot kg^{-1}}$$

$$\times 0.07143 \times \frac{1}{16}\ \mathrm{m}$$

$$= \mathbf{0.03066\ Pa}$$

Chapter 9

9.1 $\dfrac{\partial u}{\partial x} + \dfrac{\partial v}{\partial y} = 0 \quad \therefore$ Continuity is satisfied

But $\dfrac{\partial v}{\partial x} - \dfrac{\partial u}{\partial y} \neq 0 \quad \therefore$ Flow is not irrotational

9.2 Only (a) and (e) satisfy eqn 9.9

9.3
$$u = -\frac{\partial \psi}{\partial y} = 4y = -\frac{\partial \phi}{\partial x} \quad \therefore \phi = -4xy + f(y)$$

$$v = \frac{\partial \psi}{\partial x} = 1 + 4x = -\frac{\partial \phi}{\partial y} \quad \therefore \phi = -y - 4xy + f(x)$$

$\therefore$ $\boldsymbol{\phi = -y - 4xy + \textbf{const.}}$

$$\zeta = \frac{\partial v}{\partial x} - \frac{\partial u}{\partial y} = 4 - 4 = 0 \quad \therefore \textbf{Flow irrotational}$$

At $(1, -2)$ $\quad q^2 = u^2 + v^2 = (64 + 25)\ \text{m}^2 \cdot \text{s}^{-2} = 89\ \text{m}^2 \cdot \text{s}^{-2}$

$$\therefore p_0^* = \left(4800 + \frac{1}{2} \times 1.12 \times 89\right) \text{Pa}$$

At $(9, 6)$ $\quad q^2 = u^2 + v^2 = (576 + 1369)\ \text{m}^2 \cdot \text{s}^{-2} = 1945\ \text{m}^2 \cdot \text{s}^{-2}$

$$\therefore p^* = p_0^* - \frac{1}{2}\varrho q^2 = \left\{4800 + \frac{1}{2} \times 1.12(89 - 1945)\right\} \text{Pa}$$

$$= \mathbf{3761\ Pa}$$

9.4
$$q_r = -\frac{\partial \phi}{\partial r} = -\frac{A}{r} = -\frac{1}{r}\frac{\partial \psi}{\partial \theta} \quad \therefore \frac{\partial \psi}{\partial \theta} = A \text{ and } \psi = A\theta + f(r)$$

$$q_t = -\frac{\partial \phi}{r\partial \theta} = 0 = \frac{\partial \psi}{\partial r} \quad \therefore \psi = f(\theta)$$

$\therefore$ $\boldsymbol{\psi = A\theta + \textbf{const.}}$

9.5 $r = (1.5^2 + 2^2)^{1/2}\text{m} = 2.5\ \text{m} \quad \therefore q_r = \dfrac{3\pi/2}{2\pi 2.5}\ \text{m}\cdot\text{s}^{-1} = \mathbf{0.3\ m\cdot s^{-1}}$

Along radial streamlines $p^* + \dfrac{1}{2}\varrho q_r^2 = \text{constant}$

$$\therefore \frac{\partial p^*}{\partial r} = -\varrho q_r \frac{\partial q_r}{\partial r} = -\varrho q_r\left(-\frac{m}{2\pi r^2}\right)$$

$$= 800\ \text{kg}\cdot\text{m}^{-3} \times 0.3\ \text{m}\cdot\text{s}^{-1}\frac{3\pi/2}{2\pi(2.5\ \text{m})^2}\ \text{m}^2\cdot\text{s}^{-2}$$

$$= \mathbf{28.8\ Pa\cdot m^{-1}}$$

$$\text{Acceleration} = q_r\frac{\partial q_r}{\partial r} = -0.3\frac{3\pi/2}{2\pi 2.5^2}\ \text{m}\cdot\text{s}^{-2} = \mathbf{-0.036\ m\cdot s^{-2}}$$

9.6 $\dfrac{dh}{dr} = \dfrac{u^2}{gr} \quad \text{and} \quad ur = \text{constant} = C \quad \therefore \dfrac{dh}{dr} = \dfrac{C^2}{gr^3}$

$$\therefore \Delta h = \frac{C^2}{g}\int_{r_1}^{r_2}\frac{\text{dr}}{r^3} = \frac{C^2}{2g}\left[\frac{1}{r_1^2} - \frac{1}{r_2^2}\right]$$

$$Q = \int_{r_1}^{r_2} s\,u\text{dr} = sC\int_{r_1}^{r_2}\frac{dr}{r} = sC\ln\left(\frac{r_2}{r_1}\right)$$

$$= s\left\{\frac{2g(\Delta h)r_1^2 r_2^2}{r_2^2 - r_1^2}\right\}^{1/2}\ln\left(\frac{r_2}{r_1}\right) = sr_1r_2\left\{\frac{2g(\Delta h)}{r_2^2 - r_1^2}\right\}^{1/2}\ln\left(\frac{r_2}{r_1}\right)$$

$$= s\left(r + \frac{s}{2}\right)\left(r - \frac{s}{2}\right)$$

$$\times\left\{\frac{2g(\Delta h)}{\left(r + \frac{s}{2}\right)^2 - \left(r - \frac{s}{2}\right)^2}\right\}^{1/2}\ln\left(\frac{r + s/2}{r - s/2}\right)$$

$$= \boldsymbol{\left(r^2 - \frac{s^2}{4}\right)(sg\Delta h/r)^{1/2}\ln\left(\frac{2r + s}{2r - s}\right)}$$

9.7 $\partial p^*/\partial r = \varrho\omega^2 r$. Here $\omega = \text{const.}$,

$$\therefore p^* = \frac{1}{2}\varrho\omega^2 r^2 + \text{const.}$$

At free surface $p = 0 \quad \therefore \varrho g z = \frac{1}{2}\varrho\omega^2 r^2 + C$

(which is eqn of paraboloid)

Take $z = 0$ at vertex. Then $C = 0 \;\therefore\; z = \dfrac{\omega^2 r^2}{2g}$

$$\text{Volume of air} = \int \pi r^2 \mathrm{d}z = \int_0^R \pi r^2 \frac{\omega^2}{2g} 2r\,\mathrm{d}r = \frac{\pi\omega^2 R^4}{4g}$$

$$= \pi R^2 \left(\frac{150\text{ mm}}{3}\right)$$

$$\therefore\; \omega^2 = \frac{4g(0.05\text{ m})}{R^2}$$

$$\therefore\; \omega = \frac{(0.2 \times 9.81)^{1/2}}{0.05}\text{ rad}\cdot\text{s}^{-1} = \mathbf{28.01\ rad\cdot s^{-1}}$$

9.8 Pressure at circumference $= \dfrac{1}{2}\varrho\omega^2 R^2$

$$= \frac{1}{2} 900\text{ kg}\cdot\text{m}^{-3}\left(2\pi \times 15\text{ rad}\cdot\text{s}^{-1}\right)^2 \times (0.125\text{ m})^2$$

$$= \mathbf{62.5\ kPa}$$

$$\text{Force on elemental annulus} = \frac{1}{2}\varrho\omega^2 r^2 \times 2\pi r\,\mathrm{d}r$$

$$\therefore\; \text{Total force} = \varrho\omega^2\pi \int_0^R r^3\,\mathrm{d}r = \varrho\omega^2\pi R^4/4$$

$$= \left\{900(2\pi \times 15)^2\pi(0.125)^4/4\right\}\text{N} = \mathbf{1533\ N}$$

9.9 In forced vortex $p_2 - p_1 = \dfrac{1}{2}\varrho\omega^2\left(r_2^2 - r_1^2\right) = \dfrac{1}{2}\varrho\dfrac{q_1^2}{r_1^2}\left(r_2^2 - r_1^2\right)$

In free vortex $p_2 - p_1 = \dfrac{1}{2}\varrho\left(q_1^2 - q_2^2\right) = \dfrac{1}{2}\varrho q_1^2\left(1 - \dfrac{q_2^2}{q_1^2}\right)$

$$= \frac{1}{2}\varrho q_1^2\left(1 - \frac{r_1^2}{r_2^2}\right)$$

If $p_2 - p_1$ for forced vortex $= 2(p_2 - p_1$ for free vortex) and q_1 is same for both,

then $\dfrac{r_2^2 - r_1^2}{r_1^2} = 2\left(1 - \dfrac{r_1^2}{r_2^2}\right)$ whence $r_2^4 - 3r_1^2 r_2^2 + 2r_1^4 = 0$

$$\therefore\; r_2^2 = \frac{3r_1^2 \pm \sqrt{9r_1^4 - 8r_1^4}}{2} = 2r_1^2 \quad\text{or}\quad r_1^2$$

Reject $r_2^2 = r_1^2$ because then $p_2 - p_1 = 0$. $\therefore\; \boldsymbol{r_2 = r_1\sqrt{2}}$

9.10 For forced vortex $q = \omega r$ and $p_1^* - p_0^* = \frac{1}{2}\varrho\omega^2 r^2$

For free vortex $qr = C$ and $p^* = K - \frac{1}{2}\varrho q^2$

At interface $q = C/R = \omega R$ $\therefore C = \omega R^2$

$p^* = K - \frac{1}{2}\varrho\omega^2 R^2 = \frac{1}{2}\varrho\omega^2 R^2 + p_0^*$ $\therefore K = p_0^* + \varrho\omega^2 R^2$

At $r = \infty$, $q = 0$ $\therefore p_\infty^* = K$

$\therefore$ Difference in surface levels $= \dfrac{p_\infty^* - p_0^*}{\varrho g} = \boldsymbol{\omega^2 R^2/g}$

9.11 In forced vortex $p^* = \frac{1}{2}\varrho\omega^2 r^2 = \frac{1}{2}\varrho q^2 = \frac{1}{2}1000 \times 6^2\ \text{Pa}$

$= 18\ \text{kPa}$ where $q = 6\ \text{m}\cdot\text{s}^{-1}$

For free vortex $qr = C = \omega R^2 = \left(30\pi\ \text{rad}\cdot\text{s}^{-1}\right)(0.1\ \text{m})^2$

$= 0.3\pi\ \text{m}^2\cdot\text{s}^{-1}$

$$p^* = D - \frac{1}{2}\varrho\frac{C^2}{r^2}$$

At $R = 0.1$ m $p_{\text{forced}}^* = p_{\text{free}}^*$

that is, $\frac{1}{2}1000\,\text{kg}\cdot\text{m}^{-3}\left(30\pi\ \text{rad}\cdot\text{s}^{-1}\right)^2(0.1\ \text{m})^2$

$= D - \frac{1}{2}1000\,\text{kg}\cdot\text{m}^{-3}\dfrac{(0.3\pi\ \text{m}^2\cdot\text{s}^{-1})^2}{(0.1\ \text{m})^2}$

whence $D = 9000\pi^2$ Pa

In free vortex $q = 6\ \text{m}\cdot\text{s}^{-1}$ when

$r = \dfrac{C}{q} = \dfrac{0.3\pi}{6}\ \text{m} = \dfrac{\pi}{20}\ \text{m}$

Then $p^* = \left\{9000\pi^2 - \frac{1}{2}1000\dfrac{(0.3\pi)^2}{(\pi/20)^2}\right\}$ Pa

$= (9000\pi^2 - 18\,000)$Pa

$= 70.83$ kPa

$\therefore \Delta p^* = (70.83 - 18)\ \text{kPa} =$ **52.83 kPa**

9.12 For forced vortex $\omega = 8 \times 2\pi\ \text{rad}\cdot\text{s}^{-1} = 50.27\ \text{rad}\cdot\text{s}^{-1}$

and $p = \frac{1}{2}\varrho\omega^2 r^2 + A$. Let $p = 0$ when $r = 0$.

$\therefore A = 0$.

Total force on top of drum caused by forced vortex

$$= \int_0^{0.15\text{m}} \frac{1}{2}\varrho\omega^2 r^2 2\pi r dr = \frac{\pi}{4}\varrho\omega^2(0.15\text{ m})^4$$

$$= \frac{\pi}{4} 900\,\text{kg}\cdot\text{m}^{-3}\left(50.27\,\text{rad}\cdot\text{s}^{-1}\right)^2(0.15\text{ m})^4 = \mathbf{904\ N}$$

For free vortex $q_r = C$ and $p = D - \frac{1}{2}\varrho q^2$

At $r = 0.15$ m, $p = \frac{1}{2}\varrho\omega^2(0.15\text{ m})^2$ and $q = \omega(0.15\text{ m})$ as for forced vortex

$$\therefore \frac{1}{2}\varrho\omega^2(0.15\text{ m})^2 = D - \frac{1}{2}\varrho\omega^2(0.15\text{ m})^2$$

whence $D = \varrho\omega^2(0.15\text{ m})^2 = 900 \times 50.27^2 \times 0.15^2$ Pa

$= 51.2$ kPa

$C = qr = (50.27 \times 0.15)0.15\text{ m}^2\cdot\text{s}^{-1} = 1.131\text{ m}^2\cdot\text{s}^{-1}$

$$\therefore \text{Total force for free vortex } = \int_{0.15\text{m}}^{0.3\text{m}} \left(D - \frac{1}{2}\varrho\frac{C^2}{r^2}\right) 2\pi r\,dr$$

$$= 2\pi\left[\frac{Dr^2}{2} - \frac{1}{2}\varrho C^2 \ln\frac{r}{r_0}\right]_{0.15\text{m}}^{0.3\text{m}}$$

(where $\ln r_0$ = integration constant)

$$= 2\pi\left[\frac{51\,200}{2}(0.3^2 - 0.15^2) - \frac{1}{2}900 \times 1.131^2 \ln 2\right]\text{N}$$

$= 8343$ N

$\therefore$ Grand total $= 904\text{ N} + 8343\text{ N} \approx \mathbf{9250\ N}$

9.13 For a sink $\psi = -\dfrac{\theta}{2\pi}m \;\therefore \psi_1 = \dfrac{3\theta_1}{2\pi}\text{ m}^2\cdot\text{s}^{-1};\quad \psi_2 = \dfrac{2\theta_2}{\pi}\text{ m}^2\cdot\text{s}^{-1}$

$$\therefore \psi = \frac{1}{\pi}\left(\frac{3}{2}\theta_1 + 2\theta_2\right)\text{m}^2\cdot\text{s}^{-1}$$

$q_r = -\dfrac{1}{r}\dfrac{\partial\psi}{\partial\theta} = -\dfrac{3}{4\pi}\text{ m}\cdot\text{s}^{-1}$ for (1) when $r = 2$ m

and $-\dfrac{1}{\pi\sqrt{2}}\text{ m}\cdot\text{s}^{-1}$ for (2) when $r = 2\sqrt{2}$ m

$$\therefore\ u = -\frac{1}{\pi\sqrt{2}}\cos 135^\circ\ \mathrm{m\cdot s^{-1}} = \frac{1}{2\pi}\ \mathrm{m\cdot s^{-1}}$$

$$v = \left(-\frac{3}{4\pi} - \frac{1}{\pi\sqrt{2}}\cos 45^\circ\right)\mathrm{m\cdot s^{-1}} = -\frac{5}{4\pi}\ \mathrm{m\cdot s^{-1}}$$

$$\therefore\ q = (u^2+v^2)^{1/2} = \frac{1}{4\pi}\sqrt{29}\ \mathrm{m\cdot s^{-1}} = \mathbf{0.429\ m\cdot s^{-1}}$$

at $360^\circ - \arctan 2.5 = \mathbf{291.8^\circ}$

By symmetry, stagnation point is on x-axis and at $(s, 0)$ where

$$-\frac{3}{2\pi s} = -\frac{4}{2\pi(2\ \mathrm{m} - s)}\quad \therefore\ s = \frac{6}{7}\ \mathrm{m}$$

that is, stagnation point is at $\left(\mathbf{\frac{6}{7}\ m, 0}\right)$

9.14 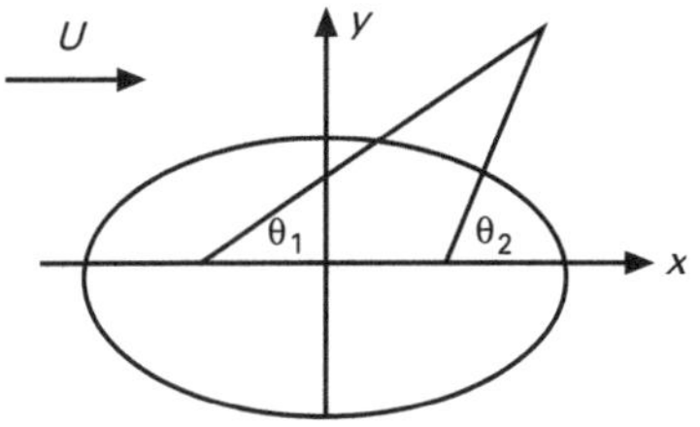

Rankine oval
Source at $(-b, 0)$, sink at $(b, 0)$, each of strength $|m|$.

Then $\psi = -UY + \dfrac{m}{2\pi}(\theta_2 - \theta_1)$

$$= -UY + \frac{m}{2\pi}\arctan\frac{2by}{x^2 - b^2 + y^2}$$

$\therefore$ Eqn of surface is: the above $= 0$

For the point $(0, Y)$ on line $\psi = 0$, $\theta_2 - \theta_1 = \dfrac{2\pi UY}{m}$

$= \pi - 2\theta_1$ by symmetry

$$\therefore\ \frac{\pi UY}{m} = \frac{\pi}{2} - \theta_1 = \arctan\frac{b}{Y}\quad \text{that is,}\ \mathbf{\frac{b}{Y} = \tan\left(\frac{\pi UY}{m}\right)}$$

Also $(-X, 0)$ is stagnation point.

$$\therefore\ 0 = U - \frac{m}{2\pi(X-b)} + \frac{m}{2\pi(X+b)}$$

whence $\mathbf{\dfrac{m}{\pi U} = \dfrac{X^2 - b^2}{b}}$

Max velocity, at $(0, Y)$, is

$$U + \frac{m}{2\pi(b^2+Y^2)^{1/2}}\cos\theta_1 + \frac{m}{2\pi(b^2+Y^2)^{1/2}}\cos(\pi - \theta_2)$$

$$= U + \frac{m}{2\pi(b^2+Y^2)^{1/2}}\left\{\frac{b}{(b^2+Y^2)^{1/2}} + \frac{b}{(b^2+Y^2)^{1/2}}\right\}$$

$$= U + \frac{mb}{\pi(b^2 + Y^2)}$$

$$= U + \frac{(X^2 - b^2)U}{b^2 + Y^2} = U\left(\frac{X^2 + Y^2}{Y^2 + b^2}\right)$$

$$\therefore \text{ Max pressure difference } = \frac{\mathbf{1}}{\mathbf{2}}\boldsymbol{\varrho U^2}\left(\frac{\mathbf{X^2 + Y^2}}{\mathbf{Y^2 + b^2}}\right)^2$$

9.15 Eliminating m from the two simultaneous eqns in Problem 9.14 gives

$$\frac{b}{Y} = \tan\frac{bY}{X^2 - b^2}$$

Then with $X = 0.1$ m and $Y = 0.05$ m, $b = 0.0781$ m (by trial)
$\therefore$ Distance between source and sink $= 2b =$ **156.2 mm**

$$|m| = \pi U\left(\frac{X^2 - b^2}{b}\right) = \pi\left(3\ \mathrm{m\cdot s^{-1}}\right)\frac{0.1^2 - 0.0781^2}{0.0781}\ \mathrm{m}$$

$$= \mathbf{0.471\ m^2\cdot s^{-1}}$$

$$\text{Max velocity } = U\left(\frac{X^2 + Y^2}{Y^2 + b^2}\right) \text{ [as in Problem 9.14]}$$

$$= \left(3\ \mathrm{m\cdot s^{-1}}\right)\left(\frac{0.1^2 + 0.05^2}{0.05^2 + 0.0781^2}\right) = \mathbf{4.36\ m\cdot s^{-1}}$$

9.16 $y_{\max} = 0.05 = \dfrac{m}{2 \times 15}$ (metre, second units)

$\therefore$ $\boldsymbol{m = 1.5\ \mathrm{m^2\cdot s^{-1}}}$

Stagnation where $r = \dfrac{m}{2\pi U} = \dfrac{1.5}{2\pi \times 15}$ m $=$ **15.92 mm**

Eqn of surface is $-UY - \dfrac{\theta m}{2\pi} = -\dfrac{m}{2}$ that is, $\theta = \pi\left(1 - \dfrac{2UY}{m}\right)$

$$\therefore \frac{y}{x} = \tan\theta = \tan\left(\pi - \frac{2\pi UY}{m}\right) = -\tan\frac{2\pi UY}{m}$$

$\therefore$ $x = -y\cot\dfrac{2\pi UY}{m} = \boldsymbol{-y\cot(20\pi y)}$ (metre units)

When $y = 0.025$ m, $x = 0$ that is, $\theta = \dfrac{\pi}{2}$

$\therefore$ $q_t = -U\sin\theta = -15\ \mathrm{m\cdot s^{-1}}$

$$q_r = U\cos\theta + \frac{m}{2\pi r} = 0 + \frac{1.5}{2\pi \times 0.025}\ \mathrm{m\cdot s^{-1}} = \frac{30}{\pi}\ \mathrm{m\cdot s^{-1}}$$

$$\Delta p = \frac{1}{2}\varrho q^2 = \frac{1}{2} \times 1.23\,\text{kg} \cdot \text{m}^{-3} \left(15^2 + \frac{30^2}{\pi^2}\right) \text{m}^2 \cdot \text{s}^{-2} = \mathbf{194.5\ Pa}$$

9.17 $\psi = -Uy - \dfrac{m}{2\pi}(\theta_1 + \theta_2); \quad \tan\theta_1 = \dfrac{y-a}{x}; \ \tan\theta_2 = \dfrac{y+a}{x}$

$$\therefore\ \tan(\theta_1 + \theta_2) = \frac{[(y-a)/x] + [(y+a)/x]}{1 - (y^2 - a^2)/x^2} = \frac{2xy}{x^2 - y^2 + a^2}$$

$$\therefore\ \boldsymbol{\psi = -Uy - \frac{m}{2\pi}\arctan\left(\frac{2xy}{x^2 - y^2 + a^2}\right)}$$

$$v = \frac{\partial \psi}{\partial x} = -\frac{m}{2\pi}\left\{1 + \frac{4x^2y^2}{(x^2 - y^2 + a^2)^2}\right\}^{-1} \times \frac{(x^2 - y^2 + a^2)\,2y - 4x^2y}{(x^2 - y^2 + a^2)^2}$$

$$= -\frac{m}{2\pi}\frac{2y(a^2 - x^2 - y^2)}{(x^2 - y^2 + a^2)^2 + 4x^2y^2}$$

$= 0$ when (i) $y = 0$ or (ii) $a^2 = x^2 + y^2$

$$u = -\frac{\partial \psi}{\partial y} = U + \frac{m}{2\pi}\left\{1 + \frac{4x^2y^2}{(x^2 - y^2 + a^2)^2}\right\}^{-1} \times \frac{(x^2 - y^2 + a^2)\,2x + 4xy^2}{(x^2 - y^2 + a^2)^2}$$

$$= U + \frac{m}{2\pi}\frac{2x(x^2 + y^2 + a^2)}{(x^2 - y^2 + a^2)^2 + 4x^2y^2}$$

For case (i): $u = U + \dfrac{mx}{\pi(x^2 + a^2)}$

$= 0$ when $\pi U x^2 + mx + \pi U a^2 = 0$

that is, when $x = -\dfrac{m}{2\pi U} \overset{(+)}{-} \sqrt{\dfrac{m^2}{4\pi^2 U^2} - a^2}$

For case (ii): $u = U + \dfrac{m}{2\pi x} = 0$ when $x = -\dfrac{m}{2\pi U}$ and

$y^2 = a^2 - \dfrac{m^2}{4\pi^2 U^2}$ which is < 0

$\therefore$ Impossible

$\therefore$ Reject case(ii)

Stagnation point is at $\left(-\dfrac{m}{2\pi U}-\sqrt{\dfrac{m^2}{4\pi^2U^2}-a^2},0\right)$

At stagnation point $\quad y=0 \quad$ and $\quad \theta_1+\theta_2=2\pi \;\therefore\; \psi=-m$

$\therefore$ Eqn of contour is $\quad Uy-m=-\dfrac{m}{2\pi}(\theta_1+\theta_2)$

When $\quad x=0,\ Uy=m-\dfrac{m}{2\pi}(\pi \text{ or } 3\pi)$ that is, $y=\pm\dfrac{m}{2U}$

$$\text{Then } u=U \quad \text{and} \quad v=-\frac{m}{2\pi}\left(\pm\frac{m}{U}\right)\left(a^2-\frac{m^2}{4U^2}\right)^{-1}$$

$$=\mp\frac{2Um^2}{\pi\left(4U^2a^2-m^2\right)}$$

$$\therefore\ q=U\left\{1+\frac{4m^4}{\pi^2(m^2-4U^2a^2)^2}\right\}^{1/2}$$

9.18 F_x (divided by depth) $=\displaystyle\int_{\pi/2}^{3\pi/2}-pa\cos\theta\ \mathrm{d}\theta$ and

$$p-p_\infty=\frac{1}{2}\varrho(U^2-q^2)=\frac{1}{2}\varrho U^2(1-4\sin^2\theta)$$

$$\therefore\ F_x=\frac{1}{2}\varrho U^2a\int_{\pi/2}^{3\pi/2}(4\sin^2\theta-1)\cos\theta\ \mathrm{d}\theta$$

$$=\frac{1}{2}\varrho U^2a\left[\frac{4}{3}\sin^3\theta-\sin\theta\right]_{\pi/2}^{3\pi/2}=-\frac{1}{3}\varrho U^2a$$

$\therefore$ Total $|F_x|=\dfrac{1}{3}\times 1000\,\mathrm{kg\cdot m^{-3}}\left(1.2\,\mathrm{m\cdot s^{-1}}\right)^2 0.9\ \mathrm{m}\times 3\ \mathrm{m}$

$=$ **1296 N**

9.19 Mass of cylinder

$=7800\,\mathrm{kg\cdot m^{-3}}\left\{\dfrac{\pi}{4}(0.3^2-0.288^2)4+2\dfrac{\pi}{4}0.288^2\times 0.006\right\}\mathrm{m^3}$

Added mass $=1000\,\mathrm{kg\cdot m^{-3}}\left(\dfrac{\pi}{4}0.3^2\times 4\right)\mathrm{m^3}=90\pi\ \mathrm{kg}$

$\therefore$ Ratio

$$=\frac{7800\left\{\pi\times 0.588\times 0.012+\pi\times 0.288^2\times 0.003\right\}+90\pi}{7800\left\{\pi\times 0.588\times 0.012+\pi\times 0.288^2\times 0.003\right\}}$$

$=$ **2.580**

9.20 $\sin\theta = \dfrac{\sqrt{3}}{2} = \dfrac{\Gamma}{4\pi aU} \quad \therefore \ |\text{Lift}| = \varrho U \dfrac{\sqrt{3}}{2} 4\pi aU$

$$\therefore \ C_L = \frac{\varrho(\sqrt{3}/2)4\pi aU^2}{\frac{1}{2}\varrho U^2 2a} = \mathbf{2\pi\sqrt{3}}$$

9.21

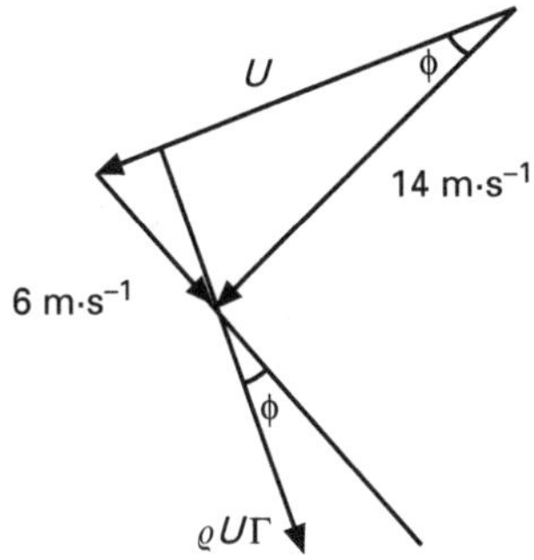

Propulsive force (SE)/
Length of cylinder

$= \varrho U\Gamma\cos\phi = \varrho\Gamma\left(14\ \text{m}\cdot\text{s}^{-1}\right)$

$\dfrac{80\times 10^3\ \text{N}}{2\times 9\ \text{m}} = 1.225\ \text{kg}\cdot\text{m}^3$

$\times\,\Gamma\times 14\ \text{m}\cdot\text{s}^{-1}$

$\therefore \ \Gamma = 259.2\ \text{m}^2\cdot\text{s}^{-1}$

But $\Gamma = 0.5\times 2\pi r\times \omega r = \pi r^2\omega$

$\therefore \ \omega = \dfrac{259.2}{\pi 1.5^2}\ \text{rad}\cdot\text{s}^{-1} = 36.67\ \text{rad}\cdot\text{s}^{-1}$ (clockwise)

$\sin\theta = \dfrac{\Gamma}{4\pi aU} = \dfrac{259.2}{4\pi 1.5\sqrt{14^2+6^2}}$

whence $\theta = 90^\circ \pm 25.49^\circ$

$\phi = \arctan 6/14 = 23.20^\circ$

Propulsion line comes from $45^\circ - 23.20^\circ = 21.8^\circ$W of N.

$\therefore$ Stagnation points are **3.69° E of N and 47.29° W of N.**
Max pressure at stagnation point. Min pressure at max velocity, which occurs at $(r = a,\ \theta = 3\pi/2)$.

$$(q_t)_{r=a} = -2U\sin\frac{3\pi}{2} + \frac{\Gamma}{2\pi a} = \left(2\sqrt{14^2+6^2} + \frac{259.2}{3\pi}\right)\text{m}\cdot\text{s}^{-1}$$

$$= 57.96\ \text{m}\cdot\text{s}^{-1}$$

Pressure difference $= \dfrac{1}{2}\varrho q^2 = \dfrac{1}{2}\times 1.225\,\text{kg}\cdot\text{m}^{-3}\left(57.96\,\text{m}\cdot\text{s}^{-1}\right)^2$

$= \mathbf{2058\ Pa}$

9.22 Flow pattern is same as right-hand half of that formed by sources at $(-a, 0)$ and $(a, 0)$

$$\therefore \ \psi = -\frac{m}{2\pi}(\theta_1+\theta_2); \quad \tan\theta_1 = \frac{y}{x+a}; \quad \tan\theta_2 = \frac{y}{x-a}$$

$$\therefore \tan(\theta_1 + \theta_2) = \frac{[y/(x+a)] + [y/(x-a)]}{1 - y^2/(x^2 - a^2)} = \frac{2xy}{x^2 - y^2 - a^2}$$

$$\therefore \psi = -\frac{m}{2\pi} \arctan \frac{2xy}{x^2 - y^2 - a^2}$$

9.23 Irrotational vortex + sink: $\psi = \frac{\Gamma}{2\pi} \ln \frac{r}{r_0} + \frac{m\theta}{2\pi}$

$$q_t = \frac{\Gamma}{2\pi r} \quad \therefore \Gamma = 2\pi r q_t = 2\pi(1.2\ \text{m})(20 \sin 70^\circ\ \text{m}\cdot\text{s}^{-1})$$

$$= 48\pi \sin 70^\circ\ \text{m}^2\cdot\text{s}^{-1}$$

$$\therefore \text{Inner } q_t = \frac{48\pi \sin 70^\circ}{2\pi 0.9}\ \text{m}\cdot\text{s}^{-1} = \frac{80}{3} \sin 70^\circ\ \text{m}\cdot\text{s}^{-1}$$

$$q_r = -\frac{m}{2\pi r} \quad \therefore m = -2\pi r q_r = -2\pi 1.2 \times 20 \cos 70^\circ\ \text{m}^2\cdot\text{s}^{-1}$$

$$= -48\pi \cos 70^\circ\ \text{m}^2\cdot\text{s}^{-1}$$

$$\therefore \text{Inner } q_r = \frac{48\pi \cos 70^\circ}{2\pi 0.9}\ \text{m}\cdot\text{s}^{-1} = \frac{80}{3} \cos 70^\circ\ \text{m}\cdot\text{s}^{-1}$$

$$\therefore q = (q_t + q_r)^{1/2} = \frac{80}{3}\ \text{m}\cdot\text{s}^{-1}$$

$$\therefore \text{Pressure drop} = \frac{1}{2} \times 1000\ \text{kg}\cdot\text{m}^3 \left\{ \left(\frac{80}{3}\right)^2 - 20^2 \right\} \text{m}^2\cdot\text{s}^{-2}$$

$$= \mathbf{155.6\ kPa}$$

9.24 $C_L \frac{1}{2} \varrho U^2 S = T \cos 7^\circ \quad \therefore C_L = \frac{102 \cos 7^\circ}{\frac{1}{2} \times 1.23 \times 13.5^2 \times 1.8 \times 0.9}$

$$= \mathbf{0.558}$$

$$C_D = \frac{102 \sin 7^\circ}{\frac{1}{2} \times 1.23 \times 13.5^2 \times 1.8 \times 0.9} = \mathbf{0.0685}$$

9.25 $C_L = \frac{32.8}{\frac{1}{2} \times 1.23 \times 30^2 \times 0.75 \times 0.1} = \mathbf{0.790}$

$$C_D = \frac{1.68}{\text{same}} = \mathbf{0.0405}$$

$$Æ = 7.5 \quad C_{Di} = \frac{0.790^2}{\pi 7.5} = \mathbf{0.02650}$$

$$\therefore C_{D\infty} = 0.0405 - 0.0265 = 0.0140$$

Effective α is less nominal by $\arctan \dfrac{0.790}{\pi 7.5} = 1.92°$.

$\therefore$ Effective $\alpha = 5.08°$

For $AR = 5$, reduction of α is $\arctan \dfrac{0.790}{\pi 5} = 2.88°$

$\therefore$ Corresponding nominal $\alpha = 5.08° + 2.88° = \mathbf{7.96°}$

$$C_L = \frac{0.790 \cos 2.88°}{\cos 1.92°} = \mathbf{0.790};$$

$$C_D = 0.0140 + \frac{0.790^2}{\pi 5} = \mathbf{0.0537}$$

Chapter 10

10.1

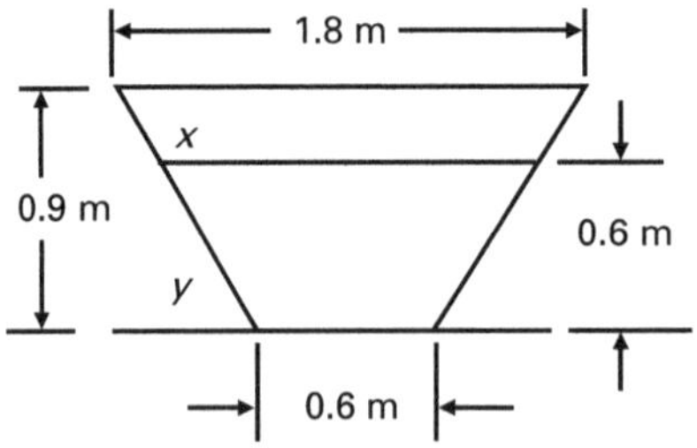

$$\frac{x}{0.6\text{ m}} = \frac{0.6}{0.9}$$

$$\therefore x = 0.4\text{ m}$$

$$y = \sqrt{0.4^2 + 0.6^2}\text{ m}$$

$$= \sqrt{0.52}\text{ m}$$

$$A = \tfrac{1}{2}\{(0.6 + 2 \times 0.4) + 0.6\}0.6\text{ m}^2 = 0.6\text{ m}^2$$

$$P = 0.6\text{ m} + 2y = 2.042\text{ m}$$

$$m = A\ P = 0.2938\text{ m}$$

$$u = 60\text{ m}^{1/2}\cdot\text{s}^{-1}\sqrt{0.2938\text{ m}\frac{1}{2600}} = 0.638\text{ m}\cdot\text{s}^{-1}$$

$$Q = 0.638\text{ m}\cdot\text{s}^{-1} \times 0.6\text{ m}^2 = \mathbf{0.3827\ m^3\cdot s^{-1}}$$

10.2

$$\frac{Q_2}{Q_1} = \frac{A_2 m_2^{2/3} i^{1/2}}{A_1 m_1^{2/3} i^{1/2}} = \left(\frac{A_2}{A_1}\right)^{5/3}\left(\frac{P_1}{P_2}\right)^{2/3}$$

$$\therefore Q_2 = 1.24\text{ m}^3\cdot\text{s}^{-1}\left(\frac{2.5 \times 0.5 + 0.5^2\tan 30^\circ}{2.5 \times 0.35 + 0.35^2\tan 30^\circ}\right)^{5/3}$$

$$\times\left(\frac{2.5 + 2 \times 0.35\operatorname{cosec} 60^\circ}{2.5 + 2 \times 0.5\operatorname{cosec} 60^\circ}\right)^{2/3} = \mathbf{2.216\ m^3\cdot s^{-1}}$$

10.3

$$n = 0.018; \quad A = (1\text{ m})\left\{b + \frac{(1\text{ m})}{1.5}\right\};$$

$$P = b + 2\sqrt{1^2 + \left(\frac{1}{1.5}\right)^2}\text{ m} = b + \frac{2}{3}\sqrt{13}\text{ m}$$

$$\text{Eqn 10.9}: \frac{2.8}{b+\frac{1}{1.5}} = \left(\frac{b+\frac{1}{1.5}}{b+\frac{2}{3}\sqrt{13}}\right)^{2/3}\left(\frac{1}{7000}\right)^{1/2} \Big/ 0.018,$$

whence (by trial) $b = \mathbf{4.456\ m}$

10.4 $n = 0.015$. For trapezium $m_{\max} = \dfrac{h}{2} = \dfrac{bh + h^2/\sqrt{3}}{b + 2 \times 2h/\sqrt{3}}$

$\therefore b = 2h\sqrt{3}$

$$0.3\ [\mathrm{m^3 \cdot s^{-1}}] = \left(bh + \frac{h^2}{\sqrt{3}}\right)\left(\frac{h}{2}\right)^{2/3}\left(\frac{1}{1800}\right)^{1/2} \Big/ 0.015$$

$$= \left(\frac{2h^2}{\sqrt{3}} + \frac{h^2}{\sqrt{3}}\right)\left(\frac{h}{2}\right)^{2/3}\left(\frac{1}{1800}\right)^{1/2} \Big/ 0.015$$

$\therefore h = \mathbf{0.520\ m}$ and $b = \mathbf{0.601\ m}$

10.5

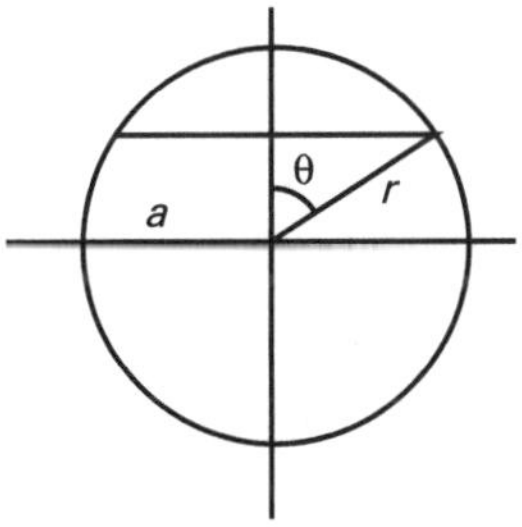

$\cos\theta = \dfrac{a}{r} = \dfrac{0.2}{0.5} \quad \therefore \theta = 66.42°$

$\therefore 2\theta = 132.84°$

$$\text{Area} = \frac{\pi}{4}(1\ \mathrm{m})^2 \frac{360 - 132.84}{360} + 0.2 \times 0.5 \sin\theta\ \mathrm{m^2} = 0.587\ \mathrm{m^2}$$

$\text{Perimeter} = \pi(1\ \mathrm{m})\dfrac{227.16}{360} = 1.982\ \mathrm{m} \quad \therefore m = 0.2962\ \mathrm{m}$

$$u = \frac{0.325}{0.587}\ \mathrm{m \cdot s^{-1}} = K(0.2962\ \mathrm{m})^{2/3}\left(\frac{1}{1500}\right)^{1/2}$$

$\therefore \mathbf{K = 48.2\ m^{1/3} \cdot s^{-1}}$

When conduit is full $i = \dfrac{4.5}{3.6 \times 10^3} = \dfrac{1}{800}$ and

$m = d/4 = 0.25\ \mathrm{m}$

$$Q = \frac{\pi}{4}(1\ \mathrm{m})^2 \times 48.2\ \mathrm{m^{1/3} \cdot s^{-1}}(0.25\ \mathrm{m})^{2/3}\left(\frac{1}{800}\right)^{1/2}$$

$= \mathbf{0.532\ m^2 \cdot s^{-1}}$

10.6

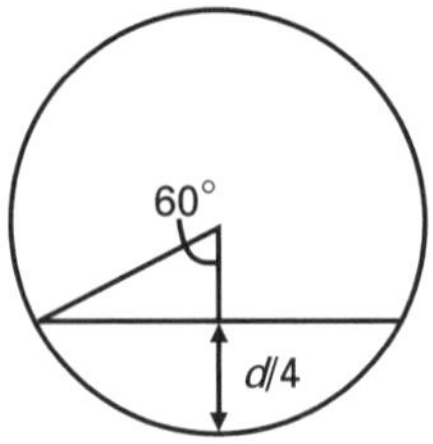

Full: $0.13\ \text{m}^3\cdot\text{s}^{-1} = \frac{\pi}{4}d^2(58\ \text{m}^{1/2}\cdot\text{s}^{-1})\sqrt{\frac{d}{4}i}$ (1)

At depth $d/4$: $A = \frac{120}{360}\frac{\pi}{4}d^2 - \frac{d}{2}\sin 60°\frac{d}{4}$

$= 0.1535d^2$

$P = \frac{120}{360}\pi d = 1.047d \quad \therefore m = 0.1466d$

$\therefore 0.6\ \text{m}\cdot\text{s}^{-1} = (58\ \text{m}^{1/2}\cdot\text{s}^{-1})\sqrt{0.1466di}$ (2)

Divide eqn (1) by eqn (2): $\frac{0.13}{0.6}\ \text{m}^2 = \frac{\pi d^2}{8\sqrt{0.1466}}$

Whence $d = \mathbf{0.460\ m}$

Then, from eqn (2), $i = \mathbf{0.001588}$

10.7

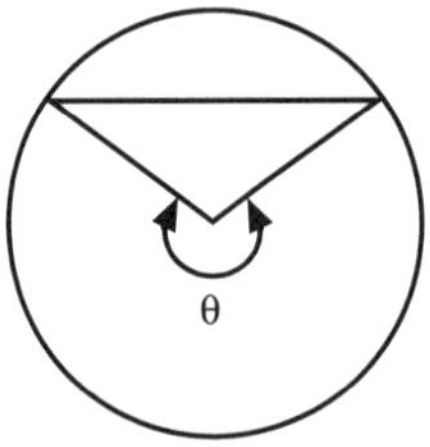

Max. Q requires max

$Am^{2/3} = A^{5/3}/P^{2/3}$

$A = \frac{1}{2}r^2\theta + r\sin\frac{\theta}{2}r\cos\left(\pi - \frac{\theta}{2}\right)$

$= r^2\left(\frac{\theta}{2} - \sin\frac{\theta}{2}\cos\frac{\theta}{2}\right)$

$= \frac{r^2}{2}(\theta - \sin\theta)$

$P = r\theta$

Max $(\theta - \sin\theta)^{5/3}/\theta^{2/3}$ requires

$\theta^{2/3}\frac{5}{3}(\theta - \sin\theta)^{2/3}(1 - \cos\theta) - (\theta - \sin\theta)^{5/3}\frac{2}{3}\theta^{-1/3} = 0$

i.e. *either*

$\theta = \sin\theta$ [which corresponds to minimum at $\theta = 0$]

or $3\theta - 5\theta\cos\theta + 2\sin\theta = 0$, whence (by trial)

$\theta = 5.278$ [in radian measure]

$\equiv \mathbf{302.4°}$

Then $A = (r^2/2)\{5.278 - (-0.844)\} = 3.061r^2$ and

$P = 5.278r$

$$2.8\ \mathrm{m^3 \cdot s^{-1}} = 3.061r^2(80\ \mathrm{m^{1/3} \cdot s^{-1}})\left(\frac{3.061\,r}{5.278}\right)^{2/3}\frac{1}{10^2},$$

whence $r = 1.205$ m

$\therefore d = \mathbf{2.41\ m}$

10.8 $A = bh + h^2/\sqrt{3} \quad \therefore A_1 = \left(1 \times 0.5 + \dfrac{0.5^2}{\sqrt{3}}\right)\mathrm{m^2} = 0.644\ \mathrm{m^2}$;

$$A_2 = \left(1 \times 0.65 + \frac{0.65^2}{\sqrt{3}}\right)\mathrm{m^2} = 0.894\ \mathrm{m^2}$$

$$u_1 = \frac{0.85}{0.644}\ \mathrm{m \cdot s^{-1}} = 1.319\ \mathrm{m \cdot s^{-1}}$$

By continuity, $A_1u_1 = A_2u_2 + (A_2 - A_1)c$

$\therefore 0.85\ \mathrm{m^3 \cdot s^{-1}} = (0.894\ \mathrm{m^2})u_2 + (0.250\ \mathrm{m^2})c$

Momentum eqn: $\varrho g A_1\bar{x}_1 - \varrho g A_2\bar{x}_2 = (u_1 + c)A_1\varrho(u_2 - u_1)$

where $\bar{x}$ = depth of centroid

that is, $\dfrac{g}{A_1}(A_1\bar{x}_1 - A_2\bar{x}_2) = (u_1 + c)(u_2 - u_1)$

$$= (u_1 + c)^2\frac{A_1 - A_2}{A_2}$$

$$A\bar{x} = bh \times \frac{h}{2} + \frac{h^2}{\sqrt{3}} \times \frac{h}{3} = \frac{bh^2}{2} + \frac{h^3}{3\sqrt{3}}$$

$$\therefore A_1\bar{x}_1 = \left(\frac{1 \times 0.5^2}{2} + \frac{0.5^3}{3\sqrt{3}}\right)\mathrm{m^3} = 0.1491\ \mathrm{m^3};$$

$$A_2\bar{x}_2 = \left(\frac{1 \times 0.65^2}{2} + \frac{0.65^3}{3\sqrt{3}}\right)\mathrm{m^3} = 0.2641\ \mathrm{m^3}$$

$$\therefore (u_1 + c)^2 = \frac{9.81}{0.644}(0.2641 - 0.1491)\frac{0.894}{0.894 - 0.644}\ \mathrm{m^2 \cdot s^{-2}}$$
$$= 6.27\ \mathrm{m^2 \cdot s^{-2}}$$

$\therefore c = (\sqrt{6.27} - 1.319)\ \mathrm{m \cdot s^{-1}} = \mathbf{1.185\ m \cdot s^{-1}}$

$Q_2 = (0.85 - 0.250 \times 1.185)\ \mathrm{m^3 \cdot s^{-1}} = \mathbf{0.554\ m^3 \cdot s^{-1}}$

Surface width $= 1\ \mathrm{m} + 2h\cot 60°$

$\therefore B_1 = 1.577$ m; $\quad B_2 = 1.751$ m

$$\therefore Fr_1 = \frac{u_1}{\sqrt{gA_1/B_1}} = \frac{1.319}{\sqrt{9.81 \times 0.644/1.577}} = \mathbf{0.659}$$

$$Fr_2 = \frac{u_2}{\sqrt{gA_2/B_2}} = \frac{0.554}{0.894}\sqrt{\frac{1.751}{9.81 \times 0.894}} = \mathbf{0.277}$$

10.9 $h + \dfrac{u^2}{2g} = 1.8\ \text{m}$ and $12\ \text{m}^3 \cdot \text{s}^{-1} = u(3\ \text{m})h$

whence $h + \dfrac{1}{2g}\left(\dfrac{4}{h}\right)^2 = 1.8$ in metre, second units

$\therefore h^3 - 1.8h^2 + 0.815 = 0 \quad \therefore h = \mathbf{1.027\ m}$ **or** $\mathbf{1.357\ m}$

$$Fr = \frac{u}{\sqrt{gh}} = \frac{4}{\sqrt{gh^3}} = \mathbf{1.226\ \text{or}\ 0.808}$$

$h_c = (q^2/g)^{1/3}$ For critical flow $\sqrt{gh_c} = m^{2/3}i^{1/2}/n$

$$\therefore i = \frac{n^2 g h_c}{m^{4/3}} = n^2 g h_c \left(\frac{b + 2h_c}{bh_c}\right)^{4/3} = \frac{n^2 (gq)^{2/3}}{b^{4/3}}\left(\frac{bg^{1/3}}{q^{2/3}} + 2\right)^{4/3}$$

$$= \frac{0.014^2 (9.81 \times 4)^{2/3}}{3^{4/3}}\left(\frac{3 \times 9.81^{1/3}}{4^{2/3}} + 2\right)^{4/3} = \mathbf{0.00394}$$

10.10 For parabola $x^2 = 4ay$. Area $= \displaystyle\int_0^y 2x\,\mathrm{d}y = 4a^{1/2}\int_0^y y^{1/2}\,\mathrm{d}y$

$$= \tfrac{8}{3}a^{1/2}y^{3/2}$$

$B = 2x = 4a^{1/2}y^{1/2} \quad \therefore A/B = \tfrac{2}{3}y$

$$\therefore u_c = \sqrt{gA/B} = \sqrt{\tfrac{2}{3}gh_c}$$

$$E = h + \frac{u^2}{2g} = h_c + \frac{u_c^2}{2g} = h_c + \frac{2}{3}\frac{gh_c}{2g} = \frac{4}{3}h_c \quad \therefore h_c = \mathbf{\frac{3}{4}E}$$

$$\therefore u_c = \sqrt{\frac{2}{3}g\frac{3}{4}E} = \sqrt{\frac{gE}{2}}$$

10.11 For min upstream depth, flow between piers is critical.

$$\therefore E = \frac{3}{2}\left(\frac{q^2}{g}\right)^{1/3} = \frac{3}{2}\left(\frac{450^2}{25^2 \times 9.81}\right)^{1/3}\ \text{m} = 4.81\ \text{m}$$

$$= h_1 + \frac{Q^2}{b_1^2 h_1^2 2g} = h_1 + \frac{450^2(\text{m}^3)}{30^2 \times 19.62 h_1^2} = h_1 + \frac{11.47(\text{m}^3)}{h_1^2}$$

whence $h_1 = \mathbf{4.15\ m}$

10.12 Wave stops when flow is critical, that is, when $Q^2B = gA^3$

$$A = bh + h^2 \cot 60° = (1 \times 0.15 + 0.15^2/\sqrt{3})\ \text{m}^2 \qquad = 0.1630\ \text{m}^2$$

$$B = b + 2h \cot 60° = (1 + 0.3/\sqrt{3})\ \text{m} \qquad = 1.173\ \text{m}$$

$$P = b + 2h \operatorname{cosec} 60° = \left(1 + 0.3 \times \frac{2}{\sqrt{3}}\right)\ \text{m} \qquad = 1.346\ \text{m}$$

$$\therefore m = \frac{A}{P} = 0.1211\ \text{m}$$

$$\frac{gA}{B} = u_c^2 = (Km^{2/3}i^{1/2})^2$$

$$\therefore K^2 = \frac{gA}{Bm^{4/3}i} = \frac{9.81 \times 0.1630}{1.173(0.1211)^{4/3}0.004}\ \text{m}^{2/3} \cdot \text{s}^{-2}$$

$$\therefore K = \mathbf{75.4\ m^{1/3} \cdot s^{-1}}$$

For uniform flow $\tau_0 = m\varrho gi = 0.1211 \times 1000 \times 9.81 \times 0.004\ \text{Pa} = \mathbf{4.75\ Pa}$

10.13 $u_1 = \dfrac{5.4}{3.5 \times 0.38}\ \text{m} \cdot \text{s}^{-1} = 4.06\ \text{m} \cdot \text{s}^{-1}$

$\therefore$ From eqn 10.25

$$h_2 = \left[-0.19 + \sqrt{0.19^2 + \frac{2 \times 0.38 \times 4.06^2}{9.81}}\right]\ \text{m} = \mathbf{0.956\ m}$$

$$\text{Head lost} = \frac{(h_2 - h_1)^3}{4h_1h_2} = \frac{0.576^3}{4 \times 0.38 \times 0.956}\ \text{m} = 0.1315\ \text{m}$$

$$\therefore \text{Power dissipated} = 5.4\ \text{m}^3 \cdot \text{s}^{-1} \times 1000\ \text{kg} \cdot \text{m}^{-3} \times 9.81\ \text{N} \cdot \text{kg}^{-1} \times 0.1315\ \text{m} = \mathbf{6970\ W}$$

10.14 For the jump $\dfrac{h_1 + h_2}{2} = \dfrac{q^2}{gh_1h_2}$

$$\therefore 1.75h_1 = \frac{(8.5/2.5)^2}{9.81 \times 2.5h_1^2}\ [\text{m}^3]$$

whence $h_1 = 0.646$ m and $h_2 = 1.615$ m

Downstream of jump $\left.\begin{aligned} A &= 2.5 \times 1.615\ \text{m}^2 = 4.04\ \text{m}^2 \\ P &= (2.5 + 2 \times 1.615)\ \text{m} = 5.73\ \text{m} \end{aligned}\right\}$

$\therefore m = 0.75$ m

$$u = \frac{8.5}{4.04}[\mathrm{m \cdot s^{-1}}] = \frac{(0.705)^{2/3}(0.002)^{1/2}}{n} \text{ whence } n = \mathbf{0.0168}$$

10.15 $$h_1 + \frac{1}{2g}\left(\frac{q}{h_1}\right)^2 = h_2 + \frac{1}{2g}\left(\frac{q}{h_2}\right)^2$$

$$\text{whence } q = h_1 h_2 \left(\frac{2g}{h_1 + h_2}\right)^{1/2} = 6 \times 1.2 \left(\frac{19.62}{7.2}\right)^{1/2} \mathrm{m^2 \cdot s^{-1}}$$
$$= 11.89 \mathrm{\ m^2 \cdot s^{-1}}$$

$\therefore Q = 6 \text{ m} \times 11.89 \mathrm{\ m^2 \cdot s^{-1}} = \mathbf{71.3\ m^3 \cdot s^{-1}}$

Critical depth $= (q^2/g)^{1/3} = (11.89^2/9.81)^{1/3}$ m $= 2.433$ m

Downstream depth > critical $\therefore$ Jump must have occurred.

Steady-flow momentum eqn:

$$\tfrac{1}{2}1000 \mathrm{\ kg \cdot m^{-3}} \times 9.81 \mathrm{\ N \cdot kg^{-1}}(1.2^2 - 3.1^2) \mathrm{\ m^2} - F$$
$$= 1000 \mathrm{\ kg \cdot m^{-3}}(11.89 \mathrm{\ m^2 \cdot s^{-1}})^2 \left(\frac{1}{3.1} - \frac{1}{1.2}\right) \mathrm{m^{-1}}$$

whence F = force on blocks divided by width = 32 080 $\mathrm{N \cdot m^{-1}}$
$\therefore$ Total force = 6 × 32 080 N = **192.5 kN**

10.16 $h_1 = (1.2 - 0.2)$ m = 1.0 m; $h_2 = 0.85$ m; $u_1 = \dfrac{q}{1.2 \text{ m}}$;

$$u_2 = \frac{q}{0.85 \text{ m}}$$

$$\therefore 1.0 + \frac{q^2}{1.2^2 \times 19.62} = 0.85 + \frac{q^2}{0.85 \times 19.62}$$

[metre, second units]

whence $q = 2.066 \mathrm{\ m^2 \cdot s^{-1}}$ and $Q = \mathbf{5.58\ m^3 \cdot s^{-1}}$

10.17 $$u_1 = \left(\frac{2.5 \times 0.9}{2.5 + 2 \times 0.9}\right)^{2/3} \left(\frac{1}{1200}\right)^{1/2} \Big/ 0.015 \mathrm{\ m \cdot s^{-1}}$$
$$= \mathbf{1.250\ m \cdot s^{-1}}$$

$$q = 1.250 \times 0.9 \mathrm{\ m^2 \cdot s^{-1}}$$
$$= 1.705 \mathrm{\ m^{1/2} \cdot s^{-1}} \left(0.9 - y + \frac{1.250^2}{19.62}\right)^{3/2} \mathrm{m^{3/2}}$$

whence $y = \mathbf{0.222\ m}$

10.18 (a) With metre, second units $0.6 + \frac{1}{19.62}\left(\frac{Q}{1.2 \times 0.6}\right)^2$

$$= 0.56 + \frac{1}{19.62}\left(\frac{Q}{0.6 \times 0.56}\right)^2$$

whence $Q = \mathbf{0.337\ m^3 \cdot s^{-1}}$

(b) $h_1 - 0.2 + \frac{1}{19.62}\left(\frac{Q}{1.2h_1}\right)^2 = \frac{3}{2}h_2$

where $h_2 = (q^2/g)^{1/3} = \left(\frac{0.337^2}{0.6^2 \times 9.81}\right)^{1/3}$ m $= 0.318$ m

whence $h_1 = 0.6674$ m

$\therefore$ Increase $= 0.0674$ m $=$ **67.4 mm**

10.19 $E = 0.43\text{ m} + \left(\frac{0.140}{0.7 \times 0.43}\right)^2 \frac{1}{19.62}\text{ m} = 0.441\text{ m}$

$\therefore$ Critical depth $= \frac{2}{3} \times 0.441$m $= 0.294$ m

Since for rectangular section, $h_c = \left(\frac{Q^2}{b^2 g}\right)^{1/3}$,

$$b = \left(\frac{Q^2}{g h_c^3}\right)^{1/2} = 0.2804\text{ m}$$

$\therefore$ Flow in throat of flume is critical and depth there $=$ **0.294 m**

$$E = 0.441\text{ m} = h_3 + \left(\frac{Q}{b_3 h_3}\right)^2 \frac{1}{2g} = h_3 + \frac{0.002039\text{ m}^3}{h_3^2}$$

i.e. $h_3^3 - 0.441h_3^2 + 0.002039 = 0$ [metre units]

But $h = 0.43$ m must fit this eqn also.

$\therefore h - 0.43$ is a factor of LHS.

$(h - 0.43)(h^2 - 0.011h - 0.00474) = 0$

whence $h_3 = (0.0055 + \sqrt{0.0055^2 + 0.00474})$ m $=$ **0.0746 m**

$$\tfrac{1}{2}\varrho g h_1 b_1 h_1 - \tfrac{1}{2}\varrho g h_3 b_3 h_3 - F = \varrho Q(u_3 - u_1)$$

$$= \varrho Q^2\left(\frac{1}{b_3 h_3} - \frac{1}{b_1 h_1}\right)$$

$$\therefore F = \varrho(h_1 - h_3)\left\{\frac{g b_1}{2}(h_1 + h_3) - \frac{Q^2}{b_1 h_1 h_3}\right\} = \mathbf{305.5\ N}$$

10.20 Near outlet $h = (q^2/g)^{1/3} = \left(\dfrac{1.25^2}{1.5^2 \times 9.81}\right)^{1/3}$ m = 0.414 m

Metre, second units:

h	*Average h*	A	P	m	u	u^2/gh	$1 - u^2/gh$	$i \times 10^4$	$(i-s)10^4$	dl/dh	Δh	Δl
0.414–0.526	0.470	0.705	2.440	0.2889	1.773	0.6818	0.3182	37.03	30.78	103.4	0.112	11.58
0.526–0.638	0.582	0.873	2.664	0.3277	1.432	0.3591	0.6409	20.42	14.17	452	0.112	50.6
0.638–0.750	0.694	1.041	2.888	0.3605	1.201	0.2118	0.7882	12.65	6.40	1232	0.112	138.0
												200.18

$i = u^2n^2/m^{4/3}$

Add about $3.5(q^2/g)^{1/3} = 1.45$ m; Total = (200.18 + 1.45) m = 201.63 m, say **202 m**

10.21 $q = \dfrac{280}{50}$ $\text{m}^2 \cdot \text{s}^{-1} = 5.6\ \text{m}^2 \cdot \text{s}^{-1}$. For uniform flow downstream

$$\frac{5.6}{h} = \frac{h^{2/3} 0.0004^{1/2}}{0.015} \quad \therefore h_2 = 2.366 \text{ m}$$

$$h_1 = \left(-\frac{2.366}{2} + \sqrt{\frac{2.366^2}{4} + \frac{2 \times 5.6^2}{9.81 \times 2.366}}\right) \text{m} = 0.842 \text{ m}$$

Metre, second units:

h	*Average h (=m)*	$u = q/h$	$u^2/gh - 1$	$i \times 10^4$	$(i-s)10^4$	dl/dh	Δh	Δl
0.6–0.681	0.6405	8.74	11.17	311.5	307.5	363.1	0.081	29.41
0.681–0.762	0.7215	7.76	7.51	209.5	205.5	365.6	0.081	29.61
0.762–0.842	0.802	6.98	5.20	147.2	143.2	362.9	0.080	29.03
								88.05 m

$$i \simeq \frac{u^2n^2}{h^{4/3}} = \frac{q^2n^2}{h^{10/3}}$$

10.22 $$c^2 = \frac{g}{m}\tanh mh = \frac{\lambda^2}{T^2} = \frac{4\pi^2}{m^2T^2} \quad \therefore m = \frac{4\pi^2}{gT^2 \tanh mh}$$

As first approximation, assume deep water $\therefore \tanh mh \simeq 1$

$\therefore m = 0.0497\ \text{m}^{-1}$

Iteration then gives $m = 0.0657\ \text{m}^{-1}$ $\therefore$ **λ = 95.6 m**
Pressure is max under crest, that is, when
$\sin mx = 1,\ \cos mx = 0$
Then (steady velocity)$^2 = [c + Am\cosh\{m(h+z)\}]^2$

$$\therefore p_{\max} = \tfrac{1}{2}\varrho[A^2m^2\sinh^2 mh + c^2]$$
$$-\tfrac{1}{2}\varrho[c + Am\cosh\{m(h+z)\}]^2 - \varrho gz$$
$$= \tfrac{1}{2}\varrho A^2m^2\sinh^2 mh - \varrho cAm\cosh\{m(h+z)\}$$
$$-\tfrac{1}{2}\varrho A^2m^2\cosh^2\{m(h+z)\} - \varrho gz$$
$$= \tfrac{1}{2}\varrho c^2a^2m^2 + \frac{\varrho c^2 am}{\sinh mh}\cosh\{m(h+z)\}$$
$$-\tfrac{1}{2}\varrho c^2a^2m^2\frac{\cosh^2\{m(h+z)\}}{\sinh^2 mh} - \varrho gz$$

Neglect terms in a^2. Then

$$p_{\max} + \varrho gz = \frac{\varrho am}{\sinh mh}\frac{g}{m}\tanh mh\cosh\{m(h+z)\}$$
$$= \varrho ag\frac{\cosh\{m(h+z)\}}{\cosh mh}$$

$$\therefore a = \left(\frac{p_{\max}}{\varrho g} + z\right)\frac{\cosh mh}{\cosh\{m(h+z)\}}$$
$$= \left(\frac{145\times 10^3}{1025\times 9.81} - 14\right)\frac{\cosh(0.0657\times 15)}{\cosh(0.0657\times 1)}\text{ m} = 0.6404\text{ m}$$

$\therefore$ Height $= 2a =$ **1.281 m**

10.23 $c = f\lambda = 25\text{ s}^{-1}\times 0.0085\text{ m} = 0.2125\text{ m}\cdot\text{s}^{-1}$

$$\therefore (0.2125\text{ m}\cdot\text{s}^{-1})^2 = \left(\frac{2\pi\gamma}{\varrho\lambda} + \frac{g\lambda}{2\pi}\right)\tanh\frac{2\pi h}{\lambda}$$
$$= \left(\frac{2\pi\gamma}{875\text{ kg}\cdot\text{m}^{-3}\times 0.0085\text{ m}} + \frac{9.81\text{ m}\cdot\text{s}^{-2}\times 0.0085\text{ m}}{2\pi}\right)\tanh\left(\frac{2\pi 0.004}{0.0085}\right)$$

whence $\gamma =$ **0.038 N·m^{-1}**

Rate of energy transmission $= \dfrac{c}{2}(1 + 2mh\operatorname{cosech} 2mh)$

$$= \frac{0.2125\text{ m}\cdot\text{s}^{-1}}{2}\left(1 + 2\frac{2\pi}{0.0085}0.004\operatorname{cosech}\frac{4\pi 0.004}{0.0085}\right)$$

$= \mathbf{0.1096\ m\cdot s^{-1}}$

$$\text{Group velocity} = \frac{c}{2}\left(\frac{3\gamma m^2 + \varrho g}{\gamma m^2 + \varrho g} + \frac{2mh}{\sinh 2mh}\right)$$
$$= \frac{\gamma m^2 c}{\gamma m^2 + \varrho g} + \text{Rate of energy transmission}$$

$$= \frac{0.038(2\pi/0.0085)^2 0.2125}{0.038(2\pi/0.0085)^2 + 875 \times 9.81} \text{ m} \cdot \text{s}^{-1}$$
$$+ 0.1096 \text{ m} \cdot \text{s}^{-1} = \mathbf{0.2600 \text{ m} \cdot \text{s}^{-1}}$$

10.24 Assume Airy waves $\therefore a = \dfrac{0.5 \text{ m}}{2} = 0.25 \text{ m}$

$$\therefore \frac{a}{h} = \frac{0.25}{4} = \frac{1}{16} : \quad \text{OK.}$$

$$\lambda = cT \quad \therefore c^2 = \left(\frac{2\pi\gamma}{\varrho cT} + \frac{gcT}{2\pi}\right) \tanh \frac{2\pi h}{cT}$$

As first approximation, assume $\tanh(2\pi h/cT) \simeq 1$
(deep water waves) and effect of γ negligible.
Then $c = gT/2\pi = (9.81 \times 5/2\pi) \text{ m} \cdot \text{s}^{-1} = 7.807 \text{ m} \cdot \text{s}^{-1}$
$\therefore \lambda = 7.807 \text{ m} \cdot \text{s}^{-1} \times 5\text{s} = 39.09 \text{ m} \gg 0.017 \text{ m}$
$\therefore \gamma$ indeed negligible

$$\text{Second approx.} : c = \left(\frac{9.81 \times 5}{2\pi} \text{ m} \cdot \text{s}^{-1}\right) \tanh \left(\frac{2\pi 4}{c5} \text{ m} \cdot \text{s}^{-1}\right)$$
$$= \left(7.807 \text{ m} \cdot \text{s}^{-1}\right) \tanh \left(\frac{1.6\pi}{c} \text{ m} \cdot \text{s}^{-1}\right)$$

By iteration $c = \mathbf{5.589 \text{ m} \cdot \text{s}^{-1}}$

$\therefore \lambda = 5.589 \text{ m} \cdot \text{s}^{-1} \times 5 \text{ s} = \mathbf{27.95 \text{ m}}$. $\quad a/\lambda < 0.01$: OK.

$m = 2\pi/\lambda = 0.2248 \text{ m}^{-1}$

$$\text{For negligible } \gamma, \quad c_g = \frac{c}{2}\left(1 + \frac{2mh}{\sinh 2mh}\right)$$
$$= \frac{5.589}{2}\left(1 + \frac{2 \times 0.2248 \times 4}{\sinh 1.7984}\right) \text{ m} \cdot \text{s}^{-1}$$
$$= \mathbf{3.407 \text{ m} \cdot \text{s}^{-1}}$$

Energy/width for one wavelength

$$= \tfrac{1}{2}a^2\lambda\varrho g = \tfrac{1}{2}(0.25 \text{ m})^2(27.95 \text{ m})1000 \text{ kg} \cdot \text{m}^{-3} \times 9.81 \text{ N} \cdot \text{kg}^{-1}$$
$$= 8570 \text{ J} \cdot \text{m}^{-1}$$

$$\therefore \text{ Total power} = 8570 \text{ J} \cdot \text{m}^{-1} \times \frac{4.5 \text{ m}}{5 \text{ s}} = \mathbf{7.71 \text{ kW}}$$

Midway between crest and trough $x = 0$ or $\lambda/2$.

At half still-water depth $z = -2$ m

$$A = -ca \operatorname{cosech} mh = -5.589 \text{ m} \cdot \text{s}^{-1} \times 0.25 \text{ m}$$
$$\times \operatorname{cosech}(0.2248 \times 4) = -1.363 \text{ m}^2 \cdot \text{s}^{-1}$$

$$
\begin{aligned}
(\text{Absolute velocity})^2 &= A^2m^2\cosh^2\{m(h+z)\}\sin^2\{m(x-ct)\} \\
&\quad + A^2m^2\sinh^2\{m(h+z)\} \\
&\quad \times\cos^2\{m(x-ct)\} \\
&= A^2m^2\sinh^2\{m(h+z)\} \quad \text{in this case} \\
\therefore |\text{Absolute velocity}| &= Am\sinh\{m(h+z)\} \\
&= 1.363\times 0.2248\sinh(0.2248\times 2)\ \text{m}\cdot\text{s}^{-1} \\
&= \mathbf{0.142\ m\cdot s^{-1}}
\end{aligned}
$$

$$
\begin{aligned}
(\text{Steady velocity})^2 &= A^2m^2\sinh^2\{m(h+z)\}\cos^2 mx \\
&\quad + [-c - Am\cosh\{m(h+z)\}\sin mx]^2
\end{aligned}
$$

$$\therefore (\text{Steady velocity at } \eta = 0)^2 = A^2m^2\sinh^2 mh + c^2$$

By Bernoulli for $x = 0$,

$$
\begin{aligned}
p &= 0 + \tfrac{1}{2}\varrho(A^2m^2\sinh^2 mh + c^2) \\
&\quad - \tfrac{1}{2}\varrho[A^2m^2\sinh^2\{m(h+z)\} + c^2] - \varrho gz \\
&= \tfrac{1}{2}\varrho A^2m^2[\sinh^2 mh - \sinh^2\{m(h+z)\}] + \varrho g(-z) \\
&= \tfrac{1}{2}1000\times 1.363^2\times 0.2248^2[\sinh^2(0.2248\times 4) \\
&\quad - \sinh^2(0.2248\times 2)] + 1000\times 9.81\times 2 \quad (\text{Pa}) \\
&= \mathbf{19.66\ kPa}
\end{aligned}
$$

$$
\begin{aligned}
\text{Horizontal semi-axis} &= \left|\frac{A}{c}\cosh\{m(h+\bar{z})\}\right| \\
&= \frac{1.363}{5.589}\cosh(0.2248\times 2)\ \text{m} = \mathbf{0.2688\ m} \\
\text{Vertical semi-axis} &= \left|\frac{A}{c}\sinh\{m(h+\bar{z})\}\right| = \mathbf{0.1136\ m}
\end{aligned}
$$

10.25 Mean rate of energy transfer/width

$$= \tfrac{1}{4}ca^2\varrho g(1 + 2mh\,\text{cosech}\,2mh) = \tfrac{1}{4}ca^2\varrho g \quad \text{for deep water.}$$

It remains constant as waves approach shore.

$$\frac{\lambda^2}{T^2} = c^2 = \frac{g}{m} = \frac{g\lambda}{2\pi} \text{ for deep water.}$$

$$\therefore \lambda = \frac{gT^2}{2\pi} \quad \text{and} \quad c = \frac{gT}{2\pi}$$

$$\therefore \text{Power/width} = \frac{a^2 \varrho g^2 T}{8\pi}$$

$$= \frac{0.6^2 \times 1025 \times 9.81^2 \times 8}{8\pi} \text{ W} \cdot \text{m}^{-1}$$

$$= 11\,300 \text{ W} \cdot \text{m}^{-1}$$

$$\therefore \text{ Total power produced} = 11\,300 \text{ W} \cdot \text{m}^{-1} \times 80 \text{ m} \times 0.5$$

$$= 452 \textbf{ kW}$$

$$\lambda^2 = \frac{4\pi^2}{m^2} = c^2 T^2 = \frac{g}{m} T^2 \tanh mh \quad \therefore m = \frac{4\pi^2}{gT^2} \coth mh$$

$$\therefore \text{When} \quad h = 5 \text{ m} \quad m = \frac{4\pi^2}{9.81 \times 8^2} \coth(m5)$$

[metre units]
whence $m = 0.1184 \text{ m}^{-1}$

$$11\,300 \text{ W} \cdot \text{m}^{-1} = \tfrac{1}{4} c a^2 \varrho g (1 + 2\, mh \operatorname{cosech} 2mh)$$

$$= \frac{1}{4} \left(\frac{2\pi}{mT} \right) a^2 \varrho g (1 + 2mh \operatorname{cosech} 2mh)$$

Hence $a = \mathbf{0.614\ m}$

Chapter 11

11.1 $$\frac{T_2}{288.15\text{ K}} = \frac{1}{1+(\gamma-1)/2} = \frac{1}{1.2} \quad \therefore T_2 = 240.125\text{ K}$$

$$\therefore u_2 = a = \sqrt{(1.4 \times 287\text{ J}\cdot\text{kg}^{-1}\cdot\text{K}^{-1} \times 240.125\text{ K})}$$
$$= \mathbf{310.6\ m\cdot s^{-1}}$$

$$\varrho_2 = \varrho_1\left(\frac{T_2}{T_1}\right)^{1/(\gamma-1)} = \frac{p_1}{RT_1}\left(\frac{1}{1.2}\right)^{2.5}$$

$$= \frac{101.3 \times 10^3}{287 \times 288.15 \times 1.2^{2.5}}\text{ kg}\cdot\text{m}^{-3} = \mathbf{0.777\ kg\cdot m^{-3}}$$

Max u when $p = 0$. $$\frac{\gamma}{\gamma-1}\frac{p_0}{\varrho_0} = \frac{u^2}{2} + 0$$

$$\therefore u_{\max} = \sqrt{\frac{2\gamma RT_0}{\gamma-1}} = \sqrt{\frac{2 \times 1.4 \times 287 \times 288.15}{0.4}}\text{ m}\cdot\text{s}^{-1}$$
$$= \mathbf{761\ m\cdot s^{-1}}$$

11.2 $$\frac{T_2}{288.15\text{ K}} = \frac{1+0.2\times 0^2}{1+0.2\times 1.6^2} \quad \therefore T_2 = 190.6\text{ K}$$

$$\therefore a_2 = \sqrt{1.4 \times 287 \times 190.6}\text{ m}\cdot\text{s}^{-1} = 276.7\text{ m}\cdot\text{s}^{-1}$$

$$\therefore u_2 = 1.6 \times 276.7\text{ m}\cdot\text{s}^{-1} = \mathbf{443\ m\cdot s^{-1}}$$

$$\Big[\text{or } u_2^2 = 2c_p(T_0 - T_2) = 2010\text{ J}\cdot\text{kg}^{-1}\cdot\text{K}^{-1}(288.15 - 190.6)\text{ K}$$

$$\text{whence } u_2 = 443\text{ m}\cdot\text{s}^{-1}\Big]$$

$$\left(\frac{p_2}{101.5\text{ kPa}}\right)^{0.4/1.4} = \frac{T_2}{T_0} = \frac{1}{1.512} \quad \therefore p_2 = \mathbf{23.88\ kPa}$$

$$\varrho_2 = \frac{p_2}{RT_2} = \frac{23.88 \times 10^3}{287 \times 190.6}\text{ kg}\cdot\text{m}^{-3} = 0.437\text{ kg}\cdot\text{m}^{-3}$$

$$\therefore \dot{m} = 0.437\ \text{kg}\cdot\text{m}^{-3}(0.6\ \text{m})^2 \times 443\ \text{m}\cdot\text{s}^{-1} = \mathbf{69.6\ kg\cdot s^{-1}}$$

Min area is where $M = 1$. From eqn 11.59

$$A_t = \frac{\dot{m}}{\{p_0\varrho_0\gamma[2/(\gamma+1)]^{(\gamma+1)/(\gamma-1)}\}^{1/2}}$$
$$= \frac{69.6}{\{[(101.5\times10^3)^2/(287\times288.15)]1.4(2/2.4)^6\}^{1/2}}\ \text{m}^2$$
$$= \mathbf{0.288\ m^2}$$

11.3 $M = \text{cosec}40°$. $a = \sqrt{1.4\times287\times283.15}\ \text{m}\cdot\text{s}^{-1} = 337.3\ \text{m}\cdot\text{s}^{-1}$

$$\text{Temperature rise} = \frac{u^2}{2c_p} = \frac{a^2M^2}{2c_p} = \frac{337.3^2\,\text{cosec}^2 40°}{2010}\ \text{K}$$
$$= 137\ \text{K}$$
$$\therefore \text{Temperature} = \mathbf{147\ °C}$$

11.4 Using eqns 11.32, 11.24 and 11.29

$$3 = \frac{(p_0)_2}{p_1} = \frac{(p_0)_2}{p_2}\frac{p_2}{p_1} = \left(1+0.2M_2^2\right)^{3.5}\frac{1+1.4M_1^2}{1+1.4M_2^2}$$
$$= \left\{1+0.2\left(\frac{1+0.2M_1^2}{1.4M_1^2-0.2}\right)\right\}^{3.5}$$
$$\times\frac{1+1.4M_1^2}{1+1.4\left[(1+0.2M_1^2)/(1.4M_1^2-0.2)\right]}$$

whence $M_1 = \mathbf{1.386}$

$$\frac{p_2}{p_1} = \frac{(1+1.4M_1^2)(1.4M_1^2-0.2)}{1.68M_1^2+1.2} = \frac{1.4M_1^2-0.2}{1.2} = \frac{7M_1^2-1}{6}$$

From eqn 11.30: $\dfrac{\varrho_2}{\varrho_1} = \dfrac{0.4+2.4p_2/p_1}{2.4+0.4p_2/p_1} = \dfrac{6M_1^2}{M_1^2+5} = \mathbf{1.665}$

Eqn 11.26: $\dfrac{T_2}{283.15\ \text{K}} = \dfrac{1+0.2M_1^2}{1+0.2M_2^2}$

$$M_2^2 = \frac{1+0.2M_1^2}{1.4M_1^2-0.2}\quad\therefore M_2 = 0.746\quad\therefore T_2 = 352.7\ \text{K}$$

$$\therefore a_2 = \sqrt{1.4\times287\times352.7}\ \text{m}\cdot\text{s}^{-1} = 376.4\ \text{m}\cdot\text{s}^{-1}$$

$$\therefore u_2 = a_2M_2 = \mathbf{280.7\ m\cdot s^{-1}}$$

For subsequent expansion $T_3 = T_2(p_1/p_2)^{0.4/1.4}$

$$= 352.7\left(\frac{6}{7M_1^2 - 1}\right)^{1/3.5} \text{K} = 286.3\ \text{K} = \mathbf{13.2\ °C}$$

11.5 $\beta_1 = 40°, \beta_2 = 30°.$

From eqn 11.37 $\tan 10° = \dfrac{2\cot 40°\left(M_1^2 \sin^2 40° - 1\right)}{M_1^2\left(1.4 + \cos 80°\right) + 2}$

whence $M_1 = \mathbf{1.967}$

Normal component $M_1 \sin\beta_1 = 1.264$

$$\frac{p_2}{p_1} = \frac{7\left(M_1 \sin\beta_1\right)^2 - 1}{6} \quad \text{[as in Problem 11.4 above]}$$

$$\therefore p_2 - p_1 = p_1\left(\frac{p_2}{p_1} - 1\right) = \frac{7}{6}p_1\left\{\left(M_1 \sin\beta_1\right)^2 - 1\right\}$$

$$= \frac{7}{6} \times 35\ \text{kPa}(1.264^2 - 1) = \mathbf{24.43\ kPa}$$

$$\left(M_2 \sin 30°\right)^2 = \frac{1 + 0.2\left(M_1 \sin 40°\right)^2}{1.4\left(M_1 \sin 40°\right)^2 - 0.2} = 0.648$$

$\therefore M_2 = 1.610$

$$\frac{303.15\ \text{K}}{T_2} = 1 + 0.2M_2^2 = 1.518 \quad \therefore T_2 = 199.7\ \text{K}$$

$$\therefore u_2 = M_2\sqrt{\gamma R T_2} = 1.610\sqrt{1.4 \times 287 \times 199.7}\ \text{m}\cdot\text{s}^{-1}$$

$$= \mathbf{456\ m\cdot s^{-1}}$$

11.6

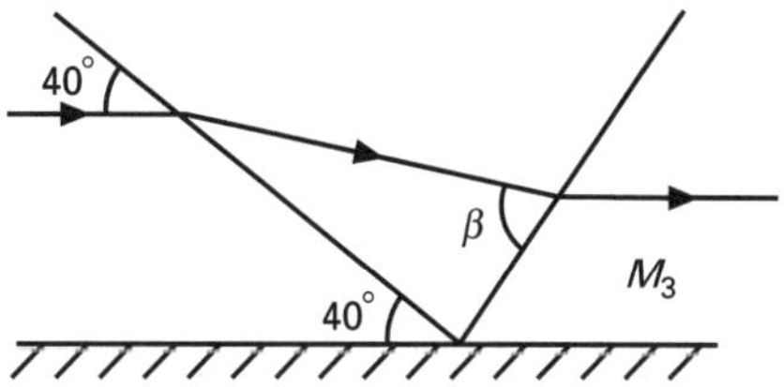

From eqn 11.37

$$\tan 10° = \frac{2\cot\beta(1.610^2 \sin^2\beta - 1)}{1.610^2\left(1.4 + \cos 2\beta\right) + 2}$$

$\therefore \beta = 50.69°$ that is, **40.69°to wall**

$$M_3^2 \sin^2 40.69° = \frac{1 + 0.2 \times 1.610^2 \sin^2 50.69°}{1.4 + 1.610^2 \sin^2 50.69° - 0.2} \quad \text{[eqn 11.29]}$$

$\therefore M_3 = \mathbf{1.250}$

11.7 $\dfrac{p_2}{p_1} = 0.5 = \left(\dfrac{1 + 0.2 \times 1.5^2}{1 + 0.2M_2^2}\right)^{1.4/0.4} \qquad \therefore M_2^2 = 3.838$

From eqn 11.42,

$$\theta_2 = \sqrt{6}\arctan\left(\frac{2.838}{6}\right)^{1/2} - \text{arcsec}(3.858)^{1/2} = 25.25^\circ$$

$$\theta_1 = \sqrt{6}\arctan\left(\frac{1.25}{6}\right)^{1/2} - \text{arcsec}1.5 = 11.91^\circ$$

$\therefore \theta_2 - \theta_1 = \mathbf{13.34^\circ}$ [or from Table A3.2]

11.8 $\dfrac{p_0}{p} > 1.893 \qquad \therefore$ Flow supersonic.

From eqn 11.47 or Table A3.1, $M_1 = 1.20$

$$T = T_0(1 + [(\gamma - 1)/2]M^2)^{-1} = 363.15\ \text{K}(1.288)^{-1} = 281.9\ \text{K}$$

$$\therefore a = \sqrt{1.4 \times 287 \times 281.9}\ \text{m} \cdot \text{s}^{-1} = 336.6\ \text{m} \cdot \text{s}^{-1}$$

$$u = 336.6 \times 1.20\ \text{m} \cdot \text{s}^{-1} = \mathbf{404\ m \cdot s^{-1}}$$

11.9 a at sea level $= \sqrt{\gamma RT} = \sqrt{1.4 \times 287 \times 288.15}\ \text{m} \cdot \text{s}^{-1}$

$$= 340.3\ \text{m} \cdot \text{s}^{-1}$$

$$\therefore \text{Assumed Mach no.} = 740 \times \frac{1000}{3600} \Big/ 340.3 = 0.604$$

This corresponds to p_2/p_1 at sea level of

$$\left(1 + [(\gamma - 1)/2]\, M^2\right)^{\gamma/(\gamma-1)} = \left\{1 + 0.2(0.604)^2\right\}^{3.5} = 1.280$$

$$\therefore p_2 - p_1 = 0.280 \times 101.3\ \text{kPa} = 28.33\ \text{kPa}$$

$$\therefore \text{Actual } p_2 = 63.83\ \text{kPa}$$

$$\frac{63.83}{35.5} = (1 + 0.2M_1^2)^{3.5} \text{ whence } M_1 = 0.955$$

a at 8000 m altitude is

$$\sqrt{1.4 \times 287 \times 236.15}\ \text{m} \cdot \text{s}^{-1} = 308.0\ \text{m} \cdot \text{s}^{-1}$$

$$\therefore u = 0.955 \times 308.0\ \text{m} \cdot \text{s}^{-1} = 294.2\ \text{m} \cdot \text{s}^{-1} = \mathbf{1059\ km \cdot h^{-1}}$$

$$T_0 - T = \frac{u^2}{2c_p} = \frac{294.2^2}{2 \times 1005}\ \text{K} = 43.07\ \text{K}$$

$$\therefore T_0 = \mathbf{6.07\ ^\circ C}$$

11.10 Critical ambient pressure = 700 kPa × 0.528 = 369.6 kPa

∴ In (a) flow does not become sonic

∴ Exit pressure = **400 kPa**

$$T = T_0\left(\frac{p}{p_0}\right)^{(\gamma-1)/\gamma} = 313.15\left(\frac{4}{7}\right)^{0.4/1.4} \text{K} = 266.9 \text{ K} \hat{=} \mathbf{-6\ °C}$$

$$u^2 = 2c_p(T_0 - T) = 2010 \times 46.25 \text{ m}^2\cdot\text{s}^{-2} \quad \therefore u = 305 \text{ m}\cdot\text{s}^{-1}$$

$$\varrho = \frac{p}{RT} = \frac{400 \times 10^3}{287 \times 266.9} \text{ kg}\cdot\text{m}^{-3} = 5.22 \text{ kg}\cdot\text{m}^{-3}$$

$$\therefore \dot{m} = \varrho A u = 5.22 \times 650 \times 10^{-6} \times 305 \text{ kg}\cdot\text{s}^{-1} = \mathbf{1.035\ kg\cdot s^{-1}}$$

(b) Exit pressure = critical = **369.6 kPa**

$$T = T_0\left(1 + \frac{\gamma-1}{2}1^2\right)^{-1} = \frac{2}{\gamma+1}T_0$$

$$= \frac{313.15}{1.2} \text{ K} = 260.96 \text{ K} \hat{=} \mathbf{-12.2\ °C}$$

$$u = \sqrt{\gamma RT} = \sqrt{1.4 \times 287 \times 260.96} \text{ m}\cdot\text{s}^{-1} = 323.8 \text{ m}\cdot\text{s}^{-1}$$

$$\varrho = \frac{369.6 \times 10^3}{287 \times 260.96} \text{ kg}\cdot\text{m}^{-3} = 4.94 \text{ kg}\cdot\text{m}^{-3}$$

$$\therefore \dot{m} = 4.94 \times 650 \times 10^{-6} \times 323.8 \text{ kg}\cdot\text{s}^{-1} = \mathbf{1.039\ kg\cdot s^{-1}}$$

11.11 $$p_0 = p\left\{1 + \left(\frac{\gamma-1}{2}\right)M^2\right\}^{\gamma/(\gamma-1)} = 14 \text{ kPa}\left\{1 + 0.2 \times 2.8^2\right\}^{3.5}$$

$$= \mathbf{380\ kPa}$$

Pressure in throat = 0.528 × 380 kPa = **200.6 kPa**

Eqn 11.59: $$m^2 = A_t^2 p_0 \varrho_0 \gamma\left(\frac{2}{\gamma+1}\right)^{(\gamma+1)/(\gamma-1)}$$

$$= \frac{A_t^2 p_0^2}{RT_0}\gamma\left(\frac{2}{\gamma+1}\right)^{(\gamma+1)/(\gamma-1)}$$

$$\therefore A_t = \frac{m\sqrt{RT_0}}{p_0\sqrt{\gamma}}\left\{\left(\frac{\gamma+1}{2}\right)^{(\gamma+1)/(\gamma-1)}\right\}^{1/2}$$

$$= \frac{1.2\sqrt{287 \times 293.15}}{380 \times 10^3\sqrt{1.4}} \times 1.2^3 \text{ m}^2$$

$$= \mathbf{1.338 \times 10^{-3}\ m^2}$$

Eqn 11.59:

$$\left(\frac{A_t}{A_e}\right)^2 = \left(\frac{\gamma+1}{2}\right)^{(\gamma+1)/(\gamma-1)}$$

$$\times \frac{2}{\gamma-1}\left\{\left(\frac{p}{p_0}\right)^{2/\gamma} - \left(\frac{p}{p_0}\right)^{(\gamma+1)/\gamma}\right\}$$

$$= 1.2^6 \times 5\left\{\left(\frac{1}{2.568^{3.5}}\right)^{2/1.4} - \left(\frac{1}{2.568^{3.5}}\right)^{2.4/1.4}\right\}$$

$$= 1.2^6 \times 5\left\{\frac{1}{2.568^5} - \frac{1}{2.568^6}\right\} = \frac{2.8^2}{2.14^6}$$

$$\therefore A_e = A_t \times \frac{2.14^3}{2.8} = \mathbf{4.68 \times 10^{-3}\ m^2}$$

$$T = T_0\left(1 + \frac{\gamma-1}{2}M^2\right)^{-1} = 293.15\ \text{K}(2.568)^{-1}$$

$$= 114.15\ \text{K} \triangleq \mathbf{-159\ °C}$$

$$u = M \times a = 2.8\sqrt{1.4 \times 287 \times 114.15}\ \text{m}\cdot\text{s}^{-1} = \mathbf{600\ m\cdot s^{-1}}$$

11.12 At design conditions

$$\left(\frac{A_t}{A}\right)^2 = \left(\frac{\gamma+1}{2}\right)^{(\gamma+1)/(\gamma-1)} \frac{M^2}{(1+[(\gamma-1)/2]\,M^2)^{(\gamma+1)/(\gamma-1)}}$$

$$= 1.2^6 \frac{1.8^2}{(1+0.2\times 1.8^2)^6} = \frac{1.2^6 \times 1.8^2}{1.648^6}$$

With argon $\left(\frac{A_t}{A}\right)^2 = \left(\frac{4}{3}\right)^4 \frac{M_1^2}{\left(1+\frac{1}{3}M_1^2\right)^4}$

$$\therefore \left(\frac{4}{3}\right)^2 \frac{M_1}{\left(1+\frac{1}{3}M_1^2\right)^2} = \frac{1.2^3 \times 1.8}{1.648^3} \text{ whence } M_1 = 1.902$$

$$\frac{p_{01}}{p_1} = \left(1+[(\gamma-1)/2]\,M_1^2\right)^{\gamma/(\gamma-1)} = \left(1+\tfrac{1}{3}M_1^2\right)^{2.5} = 7.23$$

$$\frac{p_2}{p_1} = \frac{2\gamma}{\gamma+1}\left(M_1^2 - \frac{\gamma-1}{2\gamma}\right) \text{ [from eqns 11.24 and 11.29]}$$

$$= \frac{5M_1^2-1}{4} = 4.27 \quad \therefore \frac{p_{01}}{p_2} = \frac{p_{01}}{p_1}\frac{p_1}{p_2} = \frac{7.23}{4.27} = \mathbf{1.692}$$

11.13 Provided that steam remains superheated and that Boyle's law is valid then by eqn 11.58:

$$\frac{p_2}{p_0} = \left(\frac{2}{\gamma + 1}\right)^{\gamma/(\gamma-1)} = 0.5457 \;\therefore p_2 = 109.14 \text{ kPa}$$

Initial specify entropy $= 7.280 \text{ kJ}\cdot\text{kg}^{-1}\cdot\text{K}^{-1}$ (from tables) –assumed to remain constant. Steam is *not* superheated under these conditions.

$\therefore$ Limitation is that steam does not become wet at throat. Limiting p_2 for saturation (with $s = 7.280 \text{ kJ}\cdot\text{kg}^{-1}\cdot\text{K}^{-1}$) is 126.7 kPa.

Then $v_2 = \dfrac{1}{\varrho_2} = 1.359 \text{ m}^3\cdot\text{kg}^{-1}$

$$\therefore \dot{m}^2 = 2A_2^2\varrho_2^2 \frac{\gamma}{\gamma - 1}\left(\frac{p_0}{\varrho_0} - \frac{p_2}{\varrho_2}\right)$$

$$= 2\left\{\frac{\pi}{4}(0.05 \text{ m})^2\right\}^2 \frac{1}{1.359^2} \text{ kg}^2\cdot\text{m}^{-6}\frac{1.3}{0.3}$$

$$\times (200\,000 \times 0.9602 - 126\,700 \times 1.359) \text{ N}\cdot\text{m}\cdot\text{kg}^{-1}$$

$$\therefore \dot{m} = \mathbf{0.600 \text{ kg}\cdot\text{s}^{-1}}$$

Alternatively: $\frac{1}{2}u_2^2 = \Delta h_s = (2770 - 2685) \text{ kJ}\cdot\text{kg}^{-1}$

$$= 85 \times 10^3 \text{ m}^2\cdot\text{s}^{-2}$$

$$\therefore \dot{m} = u_2A_2/v_2 = \mathbf{0.596 \text{ kg}\cdot\text{s}^{-1}}$$

11.14 From eqn 11.58: $p_c = 10^6 \text{ Pa} \times \left(\dfrac{2}{2.3}\right)^{1.3/0.3} = 5.46 \times 10^5 \text{ Pa}$

Since nozzle is choked, use eqn 11.59. Apply eqn 11.60 *downstream* of shock and substitute for m^2:

$$\left(\frac{p}{p_{02}}\right)^{2/\gamma} - \left(\frac{p}{p_{02}}\right)^{(\gamma+1)/\gamma}$$

$$= \frac{\gamma - 1}{\gamma}\left(\frac{A_t}{A}\right)^2 \frac{p_{01}}{p_{02}}\frac{\varrho_{01}}{\varrho_{02}}\gamma\left(\frac{2}{\gamma + 1}\right)^{(\gamma+1)/(\gamma-1)}$$

By Boyle's law $\dfrac{\varrho_{01}}{\varrho_{02}} = \dfrac{p_{01}}{p_{02}}\dfrac{\phi(T_{02})}{\phi(T_{01})} = \dfrac{p_{01}}{p_{02}}\dfrac{\phi(h_{02})}{\phi(h_{01})} = \dfrac{p_{01}}{p_{02}}$

assuming that h depends on T only (implied by constant c_p). Also h_0 unchanged across shock.

$$\therefore \left(\frac{p}{p_{02}}\right)^{2/1.3} - \left(\frac{p}{p_{02}}\right)^{2.3/1.3}$$

$$= \frac{\gamma - 1}{2}\left(\frac{2}{\gamma + 1}\right)^{(\gamma+1)/(\gamma-1)}\left(\frac{A_t}{A}\right)^2\left(\frac{p_{01}}{p_{02}}\right)^2$$

For exit conditions

$$x^{2/1.3} - x^{2.3/1.3} = 0.15\left(\frac{2}{2.3}\right)^{2.3/0.3}(0.5)^2\left(\frac{10^6 x}{700 \times 10^3}\right)^2$$

$$= 0.0262x^2$$

where $x = p_{\text{exit}}/p_{02}$ and $p_{\text{exit}} = 700$ kPa

i.e. $1 - x^{3/13} = 0.0262x^{6/13}$

$$\therefore\ x^{3/13} = \frac{-1 \overset{+}{\underset{(-)}{}} \sqrt{1 + 0.1048}}{0.0524} = 0.975$$

$$\therefore\ x = 0.8964\ \therefore p_{02} = \frac{700}{0.8964}\ \text{kPa}$$

$$\left(\frac{p_0}{p}\right)^{(\gamma-1)/\gamma} = 1 + \frac{\gamma - 1}{2}M^2$$

$$\therefore\ M^2_{\text{exit}} = \frac{2}{\gamma - 1}\left\{\left(\frac{1}{0.8964}\right)^{0.3/1.3} - 1\right\} = 0.1704$$

$\therefore\ M_{\text{exit}} =$ **0.413**

$$\frac{p_{02}}{p_{01}} = \frac{700 \times 10^3}{0.8964} \Big/ 10^6 = 0.781$$ $\therefore$ By eqn 11.33 $M_1 =$ **1.840**

$$\text{Upstream of shock } \frac{p_0}{p} = \left\{1 + \left(\frac{\gamma - 1}{2}\right)M^2\right\}^{\gamma/\gamma-1}$$

$$= \left\{1 + 0.15 \times 1.840^2\right\}^{1.3/0.3} = 5.92$$

Eqn 11.61

$$\left(\frac{A_t}{A}\right)^2 = \left(\frac{\gamma + 1}{2}\right)^{(\gamma+1)/(\gamma-1)}\left(\frac{2}{\gamma - 1}\right)$$

$$\times \left\{\left(\frac{1}{5.92}\right)^{2/\gamma} - \left(\frac{1}{5.92}\right)^{(\gamma+1)/\gamma}\right\} = 0.424$$

$$\therefore\ \frac{A}{A_t} = \left(\frac{1}{0.424}\right)^{1/2} = \mathbf{1.535}$$

and for exit conditions,

$$0.5^2 = \left(\frac{2.3}{2}\right)^{2.3/0.3}\left(\frac{2}{0.3}\right)\left\{\left(\frac{p_{\text{exit}}}{p_0}\right)^{2/1.3} - \left(\frac{p_{\text{exit}}}{p_0}\right)^{2.3/1.3}\right\}$$

$\therefore p_{\text{exit}}/p_0 \simeq 0.940$ or 0.1063

$\therefore p_{\text{exit}} =$ **940 kPa** or **106.3 kPa** (without shocks)

11.15 $a_1 = \sqrt{1.4 \times 287 \times 323.15}\ \mathrm{m \cdot s^{-1}} = 360.3\ \mathrm{m \cdot s^{-1}}$

$$\varrho_1 = \frac{180 \times 10^3}{287 \times 323.15}\ \mathrm{kg \cdot m^{-3}} = 1.941\ \mathrm{kg \cdot m^{-3}}$$

$$u_1 = \frac{2.7\ \mathrm{kg \cdot s^{-1}}}{1.941\ \mathrm{kg \cdot m^{-3}} \times \pi/4 \times (0.1\ \mathrm{m})^2} = 177.1\ \mathrm{m \cdot s^{-1}}$$

$$\therefore M_1 = \frac{177.1}{360.3} = 0.492$$

From Table A3.3 or eqn 11.78: $f l_{\mathrm{max}} P/A = 4 f l_{\mathrm{max}}/d = 1.140$

$$\therefore l_{\mathrm{max}} = 1.140 \times \frac{0.1\ \mathrm{m}}{4 \times 0.006} = \mathbf{4.75\ m}$$

$$T_2 = T_c = \frac{T_1}{T_1/T_c} = \left\{\frac{1 + [(\gamma - 1)/2]\,M_1^2}{(\gamma + 1)/2}\right\} T_1 = 282.3\ \mathrm{K}$$

$$\simeq \mathbf{9.2\ ^\circ C}$$

$$p_2 = p_c = p_1(p_c/p_1) = p_1 M_1 \left\{\frac{1 + [(\gamma - 1)/2]\,M_1^2}{(\gamma + 1)/2}\right\}^{1/2}$$

$$= \mathbf{82.7\ kPa}$$

For mid-point $f l_{\mathrm{max}} P/A = 1.140/2 = 0.570 \quad \therefore\ M = 0.581$

$T = T_c(T/T_c) = 282.3\ \mathrm{K} \times 1.124 = 317.3\ \mathrm{K} \simeq \mathbf{44.2\ ^\circ C}$

$p = p_c \div (p_c/p) = 82.7\ \mathrm{kPa} \div 0.548 = \mathbf{150.9\ kPa}$

11.16 (a) $\varrho_1 = \dfrac{730 \times 10^3}{287 \times 303.15}\ \mathrm{kg \cdot m^{-3}} = 8.39\ \mathrm{kg \cdot m^{-3}}$

$$u_1 = \frac{2.3\ \mathrm{kg \cdot s^{-1}}}{8.39\ \mathrm{kg \cdot m^{-3}} \times \pi/4 \times (0.15\ \mathrm{m})^2} = 15.51\ \mathrm{m \cdot s^{-1}}$$

$$\therefore M_1 = \frac{15.51}{\sqrt{1.4 \times 287 \times 303.15}} = 0.0444$$

From eqn 11.78 $\left(f l_{\mathrm{max}} PA\right)_1 = 355.7$
[Interpolation in Table A3.3 not accurate for small M.]

$$\frac{flP}{A} = \frac{0.006 \times 2000\ \mathrm{m}}{0.15\ \mathrm{m}} \times 4 = 320.0$$

$\therefore (f l_{\mathrm{max}} P/A)_2 = 35.7$

$\therefore M_2 = 0.1342$ from Table A3.3 or iteration of eqn 11.78

Eqn 11.76: $$p_2 = 730\ \mathrm{kPa} \times \frac{0.0444}{0.1342}\left\{\frac{0.4 \times 0.0444^2 + 2}{0.4 \times 0.1342^2 + 2}\right\}^{1/2}$$

$$= \mathbf{241.4\ kPa}$$

(b) Eqn 11.84 : $\dfrac{(\pi/4)^2\, 0.15^4\, (730^2 - p_2^2)\, 10^6}{2 \times 2.3^2 \times 287 \times 303.15} - \ln\left(\dfrac{730}{p_2}\right)$

$= \dfrac{2 \times 0.006 \times 2000}{0.15} \quad [p_2 \text{ in kPa}]$

whence $p_2 =$ **240.8 kPa**

11.17 Eqn 11.84:

$$\frac{(\pi/4)^2 d^4 (8^2 - 6^2) 10^{10}}{2 \times 0.32^2 \times 287 \times 288.15} - \ln\frac{8}{6} = \frac{2 \times 0.006 \times 140}{d}$$

[d in metres]

i.e. $1.02 \times 10^7 d^4 - 0.2877 = 1.68/d$

As first approx. neglect log term. Then $d^5 = \dfrac{1.68}{1.02 \times 10^7}\ \mathrm{m}^5$

$\therefore d = 0.044$ m

Then $1.02 \times 10^7 d^5 = 1.68 + 0.2877 \times 0.044 = 1.693$,

whence $d =$ **0.0441 m**

11.18 $\dfrac{(\pi/4)^2 (0.075)^4 (600^2 - 240^2) 10^6}{2m^2 \times 287 \times 288.15} - \ln\left(\dfrac{600}{240}\right) = \dfrac{2 \times 0.08 \times 45}{0.075}$

whence $m = 1.842\ \mathrm{kg \cdot s^{-1}}$

$$\varrho_1 = \frac{600 \times 10^3}{287 \times 288.15}\ \mathrm{kg \cdot m^{-3}} = 7.26\ \mathrm{kg \cdot m^{-3}}$$

$$\therefore u_1 = \frac{1.842\ \mathrm{kg \cdot s^{-1}}}{7.26\ \mathrm{kg \cdot m^{-3}} \times \pi/4 \times (0.075\ \mathrm{m})^2} = 57.5\ \mathrm{m \cdot s^{-1}}$$

$$\Delta q = \frac{1}{2}\left(u_2^2 - u_1^2\right) = \frac{u_1^2}{2}\left\{\left(\frac{u_2}{u_1}\right)^2 - 1\right\} = \frac{u_1^2}{2}\left\{\left(\frac{\varrho_1}{\varrho_2}\right)^2 - 1\right\}$$

$$= \frac{u_1^2}{2}\left\{\left(\frac{p_1}{p_2}\right)^2 - 1\right\} = \frac{57.5^2}{2}(6.25 - 1)\ \mathrm{m^2 \cdot s^{-2}}$$

$= 8670\ \mathrm{m^2 \cdot s^{-2}}$

$Q = m\Delta q = 1.842\ \mathrm{kg \cdot s^{-1}} \times 8670\ \mathrm{m^2 \cdot s^{-2}} =$ **15 970 W**

11.19 $p_1/p_2 = 300/120 = 2.5$

$flP/A = 4fl/d = 4 \times 0.006 \times 85/0.05 = 40.8$

$\therefore$ From Table A3.3: $M_1 \simeq 0.12$; $M_2 \simeq 0.030$

Put inlet conditions in eqn 11.79:

$$300 \times 10^3\ \text{Pa} = \frac{m}{(\pi/4)(0.05\ \text{m})^2 0.12}\left[\frac{287 \times 288.15\ \text{m}^2 \cdot \text{s}^{-2}}{1.4\left\{1 + 0.2 \times 0.12^2\right\}}\right]^{1/2}$$

whence $m = 0.2913\ \text{kg} \cdot \text{s}^{-1}$

or put in outlet conditions:

$$120 \times 10^3 \text{Pa} = \frac{m}{(\pi/4)(0.05\ \text{m})^2 0.30}\left[\frac{287 \times 288.15\ \text{m}^2 \cdot \text{s}^{-2}}{1.4\left\{1 + 0.2 \times 0.3^2\right\}}\right]^{1/2}$$

whence $m = 0.2934\ \text{kg} \cdot \text{s}^{-1}$

$\therefore$ say $m = \mathbf{0.292\ kg \cdot s^{-1}}$

Increase by 50% gives $m = 0.438\ \text{kg} \cdot \text{s}^{-1}$

From eqn 11.80:

$$M_2^2 = \frac{1}{0.4}\left[-1 + \sqrt{\left\{1 + \frac{0.8}{1.4}\frac{0.438^2 \times 287 \times 288.15}{120^2 \times 10^6 \times (\pi/4)^2 (0.05\ \text{m})^4}\right\}}\right]$$

$$= 0.1964$$

$\therefore M_2 = 0.4332 \ \therefore \left(4fl_{\text{max}}/d\right)_2 = 1.653$

$\therefore \left(4fl_{\text{max}}/d\right)_1 = 40.8 + 1.653 = 42.453$

$\therefore M_1 = 0.1245$

$$\therefore p_1 = \frac{0.438}{(\pi/4)(0.05)^2 0.1245}\left[\frac{287 \times 288.15}{1.4\left\{1 + 0.2 \times 0.1245^2\right\}}\right]^{1/2} \text{Pa}$$

$$= \mathbf{435\ kPa}$$

Chapter 12

12.1 Values of $\frac{l}{g}\frac{du}{dt}$ are trivial.

12.2
$$u_0 = \frac{2.5}{(\pi/4)0.1^2}\ \text{m}\cdot\text{s}^{-1} = 3.183\ \text{m}\cdot\text{s}^{-1}$$

$$\frac{du}{dt} = -Kt^{5/4} \qquad \therefore\ u = -\frac{4}{9}Kt^{9/4} + u_0$$

$$\therefore\ K = \frac{u_0}{(4/9)\,8^{9/4}} = 0.0665\ [\text{m}\cdot\text{s}^{-13/4}]$$

$\therefore$ When $t = 6$ s

$$u = \left(3.183 - \tfrac{4}{9}0.0665 \times 6^{9/4}\right)\text{m}\cdot\text{s}^{-1} = \mathbf{1.517\ m\cdot s^{-1}}$$

$$h_i = -\frac{l}{g}\frac{du}{dt} = \frac{1600}{9.81}0.0665t^{5/4} = 101.9\ \text{m} \quad \text{when} \quad t = 6\ \text{s}$$

Also h_f is reduced by

$$\frac{4 \times 0.005 \times 1600}{1 \times 19.62}(3.183^2 - 1.517^2)\ \text{m} = 12.77\ \text{m}$$

$$\therefore\ H = (300 + 101.9 + 12.8)\ \text{m} = \mathbf{414.7\ m}$$

12.3
$$k = \frac{4 \times 0.005 \times 600}{1} + 10 + 2 = 24$$

$$23\ \text{m} = 24u_{\text{max}}^2/2g \qquad \therefore\ u_{\text{max}} = 4.34\ \text{m}\cdot\text{s}^{-1}$$

$$t = 12\ \text{s} = \frac{600\ \text{m}}{24 \times 4.34\ \text{m}\cdot\text{s}^{-1}}\ln\left(\frac{4.34 + u}{4.34 - u}\right)$$

whence $\mathbf{u = 3.37\ m\cdot s^{-1}}$

12.4
$$\frac{l_1}{g}\frac{du_1}{dt} + \frac{l_2}{g}\frac{du_2}{dt} + \frac{4f_1l_1}{d_1}\frac{u_1^2}{2g} + \frac{4f_2l_2}{d_2}\frac{u_2^2}{2g} = -12\ \text{m}$$

$$\therefore \frac{30}{9.81}\frac{dQ}{dt}\frac{1}{(\pi/4)(0.225)^2} + \frac{100}{9.81}\frac{dQ}{dt}\frac{1}{(\pi/4)(0.15)^2}$$

$$+ \frac{4 \times 0.007 \times 30}{0.225 \times 19.62}\frac{Q^2}{\{(\pi/4)(0.225)^2\}^2}$$

$$+ \frac{4 \times 0.008 \times 100}{0.15 \times 19.62}\frac{Q^2}{\{(\pi/4)(0.15)^2\}^2}$$

$$= -12 \quad \text{[meter, second units]}$$

$$\therefore 654\frac{dQ}{dt} + 3602Q^2 = -12$$

that is, $0.1815\dfrac{dQ}{dt} = -Q^2 - 0.003331$

$$\therefore t = \int_{0.115}^{Q} \frac{-0.1815 dQ}{Q^2 + 0.003331}$$

$$= \frac{0.1815}{\sqrt{0.003331}}\left[\arctan\frac{Q}{\sqrt{0.003331}}\right]_Q^{0.115}$$

$$= 3.144[\arctan 1.992 - \arctan(17.33Q)]$$

$\therefore$ When $t = 2$ s $\quad \dfrac{2}{3.144} = 1.106 - \arctan(17.33Q)$

whence $Q = \mathbf{0.02929\ m^3 \cdot s^{-1}}$

12.5 Load to be accelerated = (1000 + 225 + 180) kg = 1405 kg

If plunger diam. = D m, acceleration of water in pipe

$$= 1.5\left(\frac{D}{0.04}\right)^2 \text{ m} \cdot \text{s}^{-2}$$

$$= 937.5D^2 \text{ m} \cdot \text{s}^{-2}$$

Pressure drop due to this acceleration

$$= 1000 \text{ kg} \cdot \text{m}^{-3} \times 60 \text{ m} \times 937.5D^2 \text{ m} \cdot \text{s}^{-2}$$

$$= 5.625 \times 10^7 D^2 \text{ Pa}$$

Force required at plunger

$$= \{(1000 + 225 - 180)9.81 + 1130 + 1405 \times 1.5\} \text{ N} = 13\,490 \text{ N}$$

$$\therefore \frac{\pi}{4}D^2(2.75 \times 10^6 - 5.625 \times 10^7 D^2) \text{ N} = 13\,490 \text{ N}$$

$\therefore\ D^4 - 0.0489D^2 + 0.0003053 = 0$

whence $D^2 = 0.02444 \pm 0.01709$

and $D = \sqrt{0.00735} = 0.0857$ is the smaller value of D

$\therefore$ Plunger diameter = **85.7 mm**

At steady speed, force at plunger

$= \{(1000 + 225 - 180)9.81 + 1130\}\ \text{N} = 11\,380\ \text{N}$

$$\text{Pressure in cylinder} = \frac{11\,380\ \text{N}}{(\pi/4)(0.0857\ \text{m})^2} = 1.971 \times 10^6\ \text{Pa}$$

$$\text{Pressure loss in pipe} = (2.75 \times 10^6 - 1.971 \times 10^6)\ \text{Pa}$$
$$= 0.778\ \text{MPa}$$

$$\therefore h_f = \frac{0.778 \times 10^6}{1000 \times 9.81}\ \text{m} = \frac{4 \times 0.006 \times 60}{0.04} \frac{u^2}{19.62\ \text{m}\cdot\text{s}^{-2}}$$

$\therefore u = 6.58\ \text{m}\cdot\text{s}^{-1}$

$$\therefore\ \text{Lift speed} = 6.58\left(\frac{40}{85.7}\right)^2 \text{m}\cdot\text{s}^{-1} = \mathbf{1.432\ m\cdot s^{-1}}$$

12.6 Eqn 12.8: $\frac{1}{K'} = \left(\frac{1}{1.035 \times 10^9} + \frac{0.6}{0.0125 \times 207 \times 10^9}\right) \text{m}^2\cdot\text{N}^{-1}$

$\therefore K' = 8.35 \times 10^8\ \text{Pa}$

$$\therefore c = \sqrt{\frac{8.35 \times 10^8\ \text{N}\cdot\text{m}^{-2}}{880\ \text{kg}\cdot\text{m}^{-3}}} = 974\ \text{m}\cdot\text{s}^{-1}$$

$$\therefore \frac{2l}{c} = \frac{900}{974}\ \text{s} = \mathbf{0.924\ s}$$

12.7 Hoop stress $= (\Delta p)d/2t\ \therefore \Delta p = \frac{2}{d/t}\sigma_h = \frac{40 \times 10^6}{d/t}\ \text{Pa}$

Also $(\Delta p)^2 = \varrho^2 c^2 u_0^2 = \varrho^2 u_0^2 K'/\varrho = \varrho u_0^2 \left(\frac{E}{E/K + d/t}\right)$

$$\therefore \left(\frac{40 \times 10^6\ \text{Pa}}{d/t}\right)^2$$
$$= 1000\ \text{kg}\cdot\text{m}^{-3}(2\ \text{m}\cdot\text{s}^{-1})^2 \left(\frac{200 \times 10^9\ \text{Pa}}{(200 \times 10^9)/(2 \times 10^9) + d/t}\right)$$

$\therefore (d/t)^2 - 2d/t - 200 = 0$

whence $d/t = 1 \overset{+}{\underset{(-)}{}} \sqrt{201} = \mathbf{15.18}$

$$c = (\Delta p)/\varrho u_0 = \left\{ \frac{40 \times 10^6}{15.18} \Big/ (1000 \times 2) \right\} \mathrm{m \cdot s^{-1}} = 1318\ \mathrm{m \cdot s^{-1}}$$

$\therefore 2l/c = (2 \times 500/1318)\ \mathrm{s} = \mathbf{0.759\ s}$

12.8 Put: $3\ \mathrm{m \cdot s^{-1}} = B_0\sqrt{700 \times 10^3\ \mathrm{N \cdot m^{-2}}}$

$\therefore B_0 = 3.586 \times 10^{-3}\ \mathrm{m^2 \cdot N^{-1/2} \cdot s^{-1}}$

$300 \times 10^3\ \mathrm{Pa} = \Delta u\sqrt{1.24 \times 10^9\ \mathrm{N \cdot m^{-2}} \times 850\ \mathrm{kg \cdot m^{-3}}}$

whence $\Delta u = 0.2922\ \mathrm{m \cdot s^{-1}}$

Then $3.2922\ \mathrm{m \cdot s^{-1}} = B\sqrt{400 \times 10^3\ \mathrm{N \cdot m^{-2}}}$

$\therefore$ New $B = 5.205 \times 10^{-3}\ \mathrm{m^2 \cdot N^{-1/2} \cdot s^{-1}}$

$\therefore$ Increase in area $= \left(\frac{B}{B_0} - 1\right) 100\% = \mathbf{45.2\%}$

For rapid movement, time $< \frac{2l}{c} = 2l\sqrt{\frac{\varrho}{K}}$

$$= 2 \times 3000\sqrt{\frac{850}{1.24 \times 10^9}}\ \mathrm{s}$$

$$- \mathbf{4.97\ s}$$

12.9 $\frac{2l}{c} = \frac{2 \times 750}{1200}\ \mathrm{s} = 1.25\ \mathrm{s} \quad \Delta h = -\left(\frac{1200}{9.81}\ \mathrm{s}\right)\Delta u = -(122.3\ \mathrm{s})\Delta u$

Initially $3.6\ \mathrm{m \cdot s^{-1}} = B_0\sqrt{144\ \mathrm{m}} \quad \therefore B_0 = 0.3\ \mathrm{m^{1/2} \cdot s^{-1}}$

B ($\mathrm{m^{1/2} \cdot s^{-1}}$)	h (m)	u ($\mathrm{m \cdot s^{-1}}$)	Δh (m)
0.3	144	3.6	—
0.27	144 + 18.87 = **162.9**	3.446	18.87
0.24	144 − 18.87 + 42.0 = **167.1**	3.103	42.0
0.21	144 + 18.87 − 42.0 + 46.8 = **167.7**	2.720	46.8

12.10 $u = \frac{A}{a}\frac{\mathrm{d}H}{\mathrm{d}t} \quad$ Let $h_f = Cu^2 = \frac{CA^2}{a^2}\left(\frac{\mathrm{d}H}{\mathrm{d}t}\right)^2$

$$\text{Decelerating head} = H + \frac{CA^2}{a^2}\left(\frac{\mathrm{d}H}{\mathrm{d}t}\right)^2 = -\frac{l}{g}\frac{\mathrm{d}u}{\mathrm{d}t}$$

$$= -\frac{l}{g}\frac{A}{a}\frac{\mathrm{d}^2H}{\mathrm{d}t^2}$$

$$\therefore \frac{d^2H}{dt^2} + \frac{CAg}{al}\left(\frac{dH}{dt}\right)^2 + \frac{ga}{Al}H = 0$$

$$\text{Initial velocity} = \frac{42.5\ m^3 \cdot s^{-1}}{(\pi/4)(4.5\ m)^2} = 2.672\ m \cdot s^{-1}$$

$$\text{If } h_f = Cu^2 \quad \text{then} \quad C = \frac{1\ m}{(2.672\ m \cdot s^{-1})^2}$$

$$\alpha = \frac{CAg}{al} = \frac{1}{2.672^2}\left(\frac{30}{4.5}\right)^2 \frac{9.81}{730}\ m^{-1} = \mathbf{0.0836\ m^{-1}}$$

$$\beta = \frac{ga}{Al} = \frac{9.81}{730}\left(\frac{4.5}{30}\right)^2\ s^{-2} = \mathbf{3.024 \times 10^{-4}\ s^{-2}}$$

12.11 Initially $y_0 = \frac{4fl}{d}\frac{u_0^2}{2g}$ that is, $u_0^2 = \frac{gdy_0}{2fl}$

Substitute in eqn 12.36:

$$\frac{gdy_0}{2fl} = \frac{2gd}{4fl}\left(y_0 + \frac{ad}{4fA}\right) + C\exp\left(\frac{4fAy_0}{ad}\right)$$

Similarly for max height:

$$0 = \frac{2gd}{4fl}\left(y_m + \frac{ad}{4fA}\right) + C\exp\left(\frac{4fAy_m}{ad}\right)$$

$$\text{Eliminate C. Then } \frac{4f}{d}\frac{A}{a}y_m + 1 = \exp\left\{\frac{4f}{d}\frac{A}{a}(y_m - y_0)\right\}$$

$$y_0 = \frac{4fl}{d2g}\frac{Q_0^2}{[(\pi/4)d^2]^2} = \frac{32flQ_0^2}{\pi^2 gd^5} = \frac{32 \times 0.007 \times 1200 \times 1.1^2}{\pi^2 9.81 \times 0.8^5}$$

$$= 10.25\ m$$

$$\therefore (0.3418\ m^{-1})y_m + 1 = \exp\{(0.3418\ m^{-1})(y_m - 10.25\ m)\}$$

whence $y_m = -2.893$ m

$\therefore$ Total height required $=$ **17.893 m**

Chapter 13

13.1 Max. piston speed $= 2\pi 0.4 \left(\frac{0.3}{2}\right) \text{m} \cdot \text{s}^{-1} = 0.12\pi \text{ m} \cdot \text{s}^{-1}$

$\therefore$ Max. speed in pipe $= 0.12\pi \times (150/75)^2 \text{ m} \cdot \text{s}^{-1} = 0.48\pi \text{ m} \cdot \text{s}^{-1}$

Max. piston acceleration $= (2\pi 0.4)^2 \frac{0.3}{2} \text{ m} \cdot \text{s}^{-2} = 0.096\pi^2 \text{ m} \cdot \text{s}^{-2}$

$\therefore$ Max. accel. in pipe $= 0.096\pi^2 \times (150/75)^2 = 0.384\pi^2 \text{ m} \cdot \text{s}^{-2}$

$$\text{Rate of flow} = 0.98\frac{\pi}{4}(0.15)^2 \times 0.3 \times 0.4 \text{ m}^3 \cdot \text{s}^{-1}$$
$$= 6.615 \times 10^{-4}\pi \text{ m}^3 \cdot \text{s}^{-1}$$

$$\therefore \text{Mean velocity in pipes} = \frac{6.615 \times 10^{-4}\pi}{(\pi/4)(0.075)^2} \text{ m} \cdot \text{s}^{-1} = 0.4704 \text{ m} \cdot \text{s}^{-1}$$

$$\textit{Suction}: h_f \text{ at mid-stroke} = \frac{4 \times 0.01 \times 7.5}{0.075}\frac{(0.48\pi)^2}{19.62} \text{ m}$$
$$= 0.4636 \text{ m}$$

$$\text{Accel. head at ends of stroke} = \frac{7.5}{9.81}0.384\pi^2 \text{ m} = 2.897 \text{ m}$$

$$\textit{Discharge}: h_f \text{ in 15 m of pipe at mid-stroke} = \frac{15}{7.5} \times 0.4636 \text{ m}$$
$$= 0.9272 \text{ m}$$

$$\text{Steady } h_f \text{ in 285 m of pipe} = \frac{4 \times 0.01 \times 285}{0.075}\frac{0.4704^2}{19.62} \text{ m}$$
$$= 1.714 \text{ m}$$

$$\text{Accel. head in 15 m of pipe} = \frac{15}{9.81}0.384\pi^2 \text{ m} = 5.795 \text{ m}$$

$\therefore$ Head at beginning of suction stroke $= (10.33 - 3 - 2.897)$ m
$= \mathbf{4.43\ m}$

Head at middle of suction stroke $= (10.33 - 3 - 0.4636)$ m
$= \mathbf{6.87\ m}$

Head at end of suction stroke $= (10.33 - 3 + 2.897)$ m
$= \mathbf{10.23\ m}$

Head at beginning of delivery stroke
$= (10.33 + 13.5 + 1.714 + 5.795)$ m
$= \mathbf{31.34\ m}$

Head at middle of delivery stroke
$= (10.33 + 13.5 + 1.714 + 0.927)$ m
$= \mathbf{26.47\ m}$

Head at end of delivery stroke
$= (10.33 + 13.5 + 1.714 - 5.795)$ m
$= \mathbf{19.75\ m}$

13.2 Velocity in pipe $= \omega r \sin\theta \dfrac{A}{a} + \omega r \sin(90° + \theta)\dfrac{A}{a}$

$$= \frac{A}{a}\omega r(\sin\theta + \cos\theta)$$

[positive components only]

Velocity is max. at 45°, 135° etc.

$\therefore$ *Max.* velocity in pipe $= \dfrac{A}{a}\omega r\left(\dfrac{1}{\sqrt{2}} + \dfrac{1}{\sqrt{2}}\right)$

$$= \left(\frac{200}{100}\right)^2 \left(2\pi\frac{20}{60}\right) 0.225 \times \frac{2}{\sqrt{2}}\ \text{m}\cdot\text{s}^{-1}$$

$$= \mathbf{2.666\ m\cdot s^{-1}}$$

$Q = 4 \times \dfrac{\pi}{4}(0.2)^2 \times 0.45 \times \dfrac{20}{60}\ \text{m}^3\cdot\text{s}^{-1}$

$= 0.006\pi\ \text{m}^3\cdot\text{s}^{-1}$

$\therefore$ *Mean* velocity in pipe $= \dfrac{0.006\pi}{(\pi/4)(0.1)^2}\ \text{m}\cdot\text{s}^{-1} = \mathbf{2.40\ m\cdot s^{-1}}$

Inertia head is max. or min. at instant of min. velocity.

$$\text{Max}|h_\text{i}| = \frac{A}{a}\frac{l}{g}\omega^2 r = \left(\frac{200}{100}\right)^2 \frac{60}{9.81}\left(\frac{2\pi}{3}\right)^2 0.225\ \text{m} = 24.15\ \text{m}$$

$\therefore$ Inertia pressure $= \pm 1000 \times 9.81 \times 24.15\ \text{Pa} = \mathbf{\pm 236.9\ kPa}$

13.3 $$\frac{39.27P^{1/2}}{1000^{1/2}(9.81 \times 233)^{5/4}} \not> 0.138\ [\text{SI units}]$$

$\therefore P \not> 3.095 \times 10^6$ W per jet

$\therefore$ No. of jets $\not< \dfrac{30 \times 10^6}{3.095 \times 10^6} = 9.69$, say 10 that is, **5 wheels**

$$\frac{30 \times 10^6}{0.815}\ \text{W} = Q 1000\ \text{kg} \cdot \text{m}^{-3} \times 9.81\ \text{N} \cdot \text{kg}^{-1} \times 233\ \text{m}$$

$\therefore$ Total $Q = 16.10\ \text{m}^3 \cdot \text{s}^{-1}$

$\therefore Q/jet = 1.610\ \text{m}^3 \cdot \text{s}^{-1}$

Jet velocity $= 0.97\sqrt{19.62 \times 233}\ \text{m} \cdot \text{s}^{-1} = 65.6\ \text{m} \cdot \text{s}^{-1}$

$\frac{\pi}{4} d_\text{jet}^2 \times 65.6\ \text{m} \cdot \text{s}^{-1} = 1.610\ \text{m}^3 \cdot \text{s}^{-1}$ whence $d_\text{jet} = \mathbf{0.177\ m}$

Blade speed for max. efficiency $= 65.6\ \text{m} \cdot \text{s}^{-1} \times 0.46$
$= 30.17\ \text{m} \cdot \text{s}^{-1}$

$\therefore D = \dfrac{30.17}{6.25\pi}\ \text{m} = \mathbf{1.536\ m}$

$$\eta_\text{hyd} = \frac{\Delta v_\text{w} \times Q\varrho \times u}{Q\varrho g H} = \frac{u \Delta v_\text{w}}{gH}$$

$$= \frac{0.46C_\text{v}\sqrt{2gH}\left(1 - 0.85\cos 165°\right)(1 - 0.46)C_\text{v}\sqrt{2gH}}{gH}$$

$$= 0.46 \times 1.821 \times 0.54 \times 2C_\text{v}^2 = \mathbf{0.851}$$

$$v_2^2 = u^2 + R_2^2 - 2uR_2\cos(180° - \theta)$$

$$= C_\text{v}^2(2gH)\{0.46^2 + (0.85 \times 0.54)^2 - 2 \times 0.46 \times 0.85 \times 0.54\cos 15°\}$$

Remaining K.E. divided by weight $= v_2^2/2g$

$$\therefore \text{Percentage} = \frac{v_2^2}{2gH} \times 100$$

$$= 100 \times 0.97^2\{0.2116 + 0.2107 - 0.408\}$$

$$= \mathbf{1.35}$$

13.4 $u = 0.46C_v\sqrt{2gH}; \quad \omega = 2u/D$

$$Q = \frac{\pi}{4}d^2 C_v\sqrt{2gH}; \quad P = \eta Q\varrho gH = \eta\frac{\pi}{4}d^2 C_v\sqrt{2}\varrho(gH)^{3/2}$$

$$\Omega_p = \frac{\omega P^{1/2}}{\varrho^{1/2}(gH)^{5/4}}$$

$$= \frac{0.92C_v\sqrt{2}(gH)^{1/2}\eta^{1/2}\pi^{1/2}dC_v^{1/2}2^{1/4}\varrho^{1/2}(gH)^{3/4}}{D2\varrho^{1/2}(gH)^{5/4}}$$

$$= \frac{0.46C_v^{3/2}\eta^{1/2}d\pi^{1/2}2^{3/4}}{D} = \mathbf{1.21\frac{d}{D}\ rad}$$

13.5 Useful work divided by weight

$$= \frac{u}{g}(v_1 - u)(1 + k_1 \cos 15°) - k_2u^2$$

$$= \frac{u}{g}(v_1 - u)y - k_2u^2 \quad \text{where} \quad y = 1 + k_1 \cos 15°$$

$$\therefore \eta = \frac{(u/g)(v_1 - u)y - k_2u^2}{v_1^2/2g}$$

$$\therefore \frac{v_1^2}{2g}\frac{\partial\eta}{\partial u} = \frac{y}{g}(v_1 - 2u) - 2k_2u = 0$$

when $\dfrac{yv_1}{g} = 2u\left(k_2 + \dfrac{y}{g}\right)$

that is, $\dfrac{u}{v_1} = 0.46 = \dfrac{y}{2(gk_2 + y)}$ whence $gk_2 = 2y/23$

$$\text{Overall efficiency} = \left\{\frac{u}{g}(v_1 - u)y - k_2u^2\right\} \Big/ H$$

$$\therefore \text{Max. efficiency} = \left\{\frac{0.46v_1}{g}(0.54v_1)y - k_2 0.46^2 v_1^2\right\} \Big/ H$$

$$= \frac{v_1^2}{gH}(0.2484y - 0.2116gk_2)$$

$$= 2C_v^2(0.2484y - 0.2116 \times 2y/23) = 0.79$$

whence $y = 1.825 \ \therefore k_1 \cos 15° = 0.825 \ \therefore \mathbf{k_1 = 0.854}$

13.6 Total losses are $(4.8 + 8.8 + 0.79 + 0.46)$ m $= 14.85$ m

$$\therefore (120 + 3)\ \text{m} = 14.85\ \text{m} + \frac{u_1v_{w1}}{g} \quad \therefore \frac{u_1v_{w1}}{g} = 108.15\ \text{m}$$

$$\therefore v_{w1} = \frac{108.15 \times 9.81}{30}\ \text{m}\cdot\text{s}^{-1} = 35.37\ \text{m}\cdot\text{s}^{-1}$$

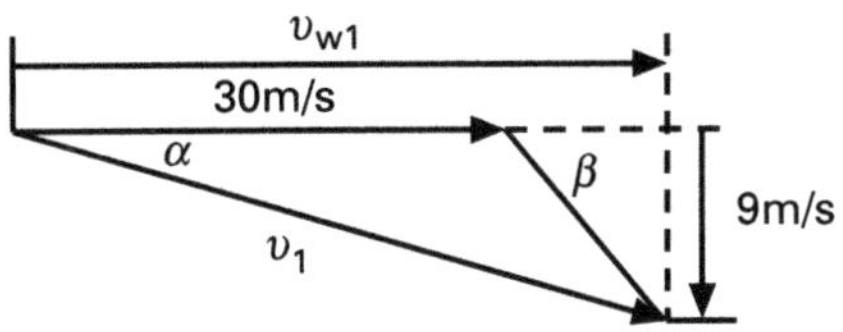

$$\alpha = \arctan\frac{9}{35.37} = \mathbf{14.28^\circ}$$

$$\beta = \arctan\frac{9}{5.37} = 59.2^\circ$$

$$\frac{v_1^2}{2g} = \frac{35.37^2 + 9^2}{19.62}\ \text{m} = 67.9\ \text{m}$$

$$(120 + 3)\ \text{m} = 4.8\ \text{m} + \frac{p_1}{\varrho g} + \frac{v_1^2}{2g} + 3\ \text{m}$$

$$\therefore \frac{p_1}{\varrho g} = (115.2 - 67.9)\ \text{m} = \mathbf{47.3\ m}$$

$$\frac{p_2}{\varrho g} + \frac{v_2^2}{2g} + 3\text{m} = (0.46 + 0.79)\ \text{m}$$

$$\therefore \frac{p_2}{\varrho g} = \left(0.46 + 0.79 - 3 - \frac{9^2}{19.62}\right)\ \text{m} = \mathbf{-5.88\ m}$$

13.7

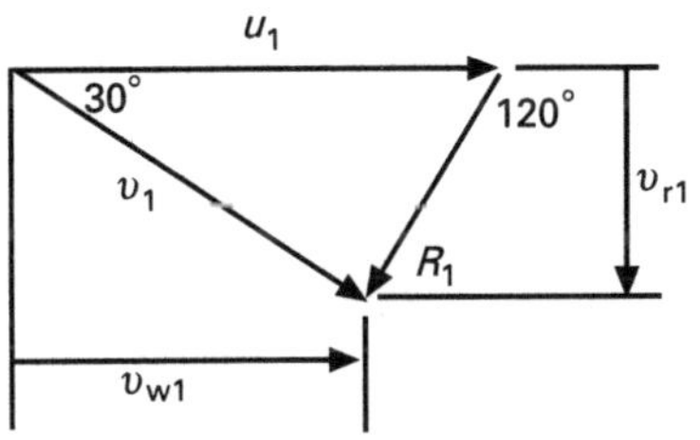

$$u_1 = \pi D_1 16.67(\text{s}^{-1}) = 52.4 D_1$$

$$v_{w1} = v_1 \cos 30^\circ = u_1 \cos^2 30^\circ = 39.28 D_1$$

$$0.88 = \frac{u v_{w1}}{gH} = \frac{52.4 D_1 39.28 D_1}{9.81 \times 15}$$

whence $D_1 = \mathbf{0.251\ m}$

$$v_{r1} = v_1 \sin 30^\circ = u_1 \cos 30^\circ \sin 30^\circ = 22.68 D_1$$

$$\therefore Q = \pi D_1 \times \frac{D_1}{4} \times v_{r1} = \frac{\pi}{4} 22.68 D_1^3 = 0.2813\ \text{m}^3 \cdot \text{s}^{-1}$$

$$\therefore P = 0.85 \times 0.2813\ \text{m}^3 \cdot \text{s}^{-1} \times 1000\ \text{kg} \cdot \text{m}^{-3} \times 9.81\ \text{N} \cdot \text{kg}^{-1} \times 15\ \text{m} = \mathbf{35\,200\ W}$$

13.8 Head at inlet $= \left(240 + \frac{8.5^2}{19.62} + 3\right)\ \text{m} = 246.7\ \text{m}$

$$\therefore \text{Power} = 15.5\ \text{m}^3 \cdot \text{s}^{-1} \times 1000\ \text{kg} \cdot \text{m}^{-3} \times 9.81\ \text{N} \cdot \text{kg}^{-1} \times 246.7\ \text{m} \times 0.9 = \mathbf{33.76\ MW}$$

$$\Omega_p = \frac{44.86(33.76 \times 10^6)^{1/2}}{1000^{1/2}(9.81 \times 246.7)^{5/4}}\ \text{rad} = \mathbf{0.486\ rad}$$

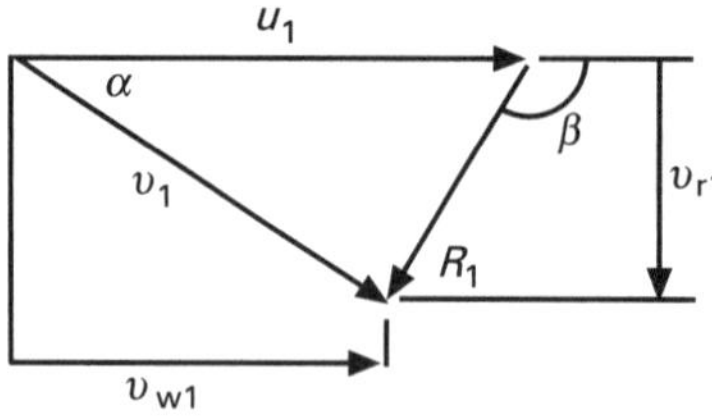

$$v_{r1} = \frac{15.5\ \text{m}^3\cdot\text{s}^{-1}}{\pi 2.23 \times 0.3\ \text{m}^2}$$
$$= 7.375\ \text{m}\cdot\text{s}^{-1}$$
$$u_1 = \frac{2.23 \times 44.86}{2}\ \text{m}\cdot\text{s}^{-1}$$
$$= 50.0\ \text{m}\cdot\text{s}^{-1}$$
$$0.93 = \eta_{\text{hyd}} = \frac{50.0 v_{w1}}{9.81 \times 246.7}\ \text{s}\cdot\text{m}^{-1}$$
$$\therefore v_{w1} = 45.0\ \text{m}\cdot\text{s}^{-1}$$

$$\therefore \alpha - \arctan\frac{7.375}{45.0} = \mathbf{9.31°};$$
$$\beta = \arctan\frac{v_{r1}}{v_{w1} - u} = \arctan\frac{7.375}{-5.0} = \mathbf{124.29°}$$

$$\frac{v_1^2}{2g} = \frac{7.375^2 + 45.0^2}{19.62}\ \text{m} = 105.9\ \text{m}$$

∴ % of net head which is kinetic at entry to runner

$$= \frac{105.9}{246.7} \times 100$$
$$= \mathbf{42.9}$$

13.9

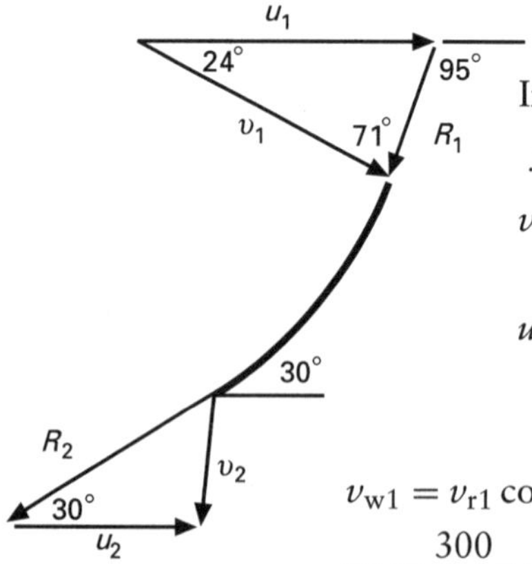

Inlet and outlet areas equal

$$\therefore v_{r1} = v_{r2}$$
$$v_1 = v_{r1}\ \text{cosec}\ 24°$$
$$u_1 = \frac{v_1 \sin 71°}{\sin 85°} = \frac{v_{r1}\sin 71°}{\sin 24° \sin 85°}$$
$$= 2.334 v_{r1}$$

$$v_{w1} = v_{r1}\cot 24°$$
$$u_2 = \frac{300}{450}u_1 = 1.556 v_{r1}$$

$$v_{w2} = u_2 - v_{r2}\cot 30°$$
$$= 1.556 v_{r1} - 1.732 v_{r1}$$
$$= -0.176 v_{r1}$$

$$v_2^2 = v_{r1}^2(1 + 0.176^2) = 1.031 v_{r1}^2$$

$$0.88 \times 55\ \mathrm{m} = \frac{u_1 v_{w1} - u_2 v_{w2}}{g} + \frac{v_2^2}{2g}$$

$$= \frac{1}{9.81\ \mathrm{m\cdot s^{-2}}}\left(2.334 \cot 24° v_{r1}^2 + 1.556 \times 0.176 v_{r1}^2 + \frac{1.031}{2} v_{r1}^2\right)$$

whence $v_{r1} = 8.87\ \mathrm{m\cdot s^{-1}}$

$$\omega = \frac{2u_1}{D} = \frac{2 \times 2.334 \times 8.87}{0.4}\ \mathrm{rad\cdot s^{-1}} = \mathbf{92\ rad\cdot s^{-1}}$$

$$Q = v_{r1} 0.92\pi 0.450 \times 0.050\ \mathrm{m^2} = 0.577\ \mathrm{m^3\cdot s^{-1}}$$

$$P = Q\varrho g\left(\frac{u_1 v_{w1} - u_2 v_{w2}}{g}\right) 0.94$$

$$= 0.577\ \mathrm{m^3\cdot s^{-1}} \times 1000\ \mathrm{kg\cdot m^{-3}}(2.334 \cot 24° + 1.556 \times 0.176) 8.87^2\ \mathrm{m^2\cdot s^{-2}} \times 0.94$$

$$= \mathbf{235.5\ kW}$$

13.10 For dynamic similarity equate $gH/\omega^2 D^2$

$$\omega_m = \omega_p \frac{D_p}{D_m}\left(\frac{H_m}{H_p}\right)^{1/2} = 44.86\ \mathrm{rad\cdot s^{-1}} \times 4\left(\frac{10.8}{30}\right)^{1/2}$$

$$= \mathbf{107.6\ rad\cdot s^{-1}}$$

$$\eta_m = \frac{100 \times 1000\ \mathrm{W}}{1.085\ \mathrm{m^3\cdot s^{-1}} \times 1000\ \mathrm{kg\cdot m^{-3}} \times 9.81\ \mathrm{N\cdot kg^{-1}} \times 10.8\ \mathrm{m}}$$

$$= 0.870 \quad \therefore \eta_p = 0.900$$

Equate $\dfrac{Q}{\omega D^3}$

$$\therefore Q_p = \frac{\omega_p}{\omega_m}\left(\frac{D_p}{D_m}\right)^3 Q_m = \frac{44.86}{107.6} \times 4^3 \times 1.085\ \mathrm{m^3\cdot s^{-1}}$$

$$= 28.93\ \mathrm{m^3\cdot s^{-1}}$$

$$\therefore P_p = 0.900 \times 28.93\ \mathrm{m^3\cdot s^{-1}} \times 1000\ \mathrm{kg\cdot m^{-3}} \times 9.81\ \mathrm{N\cdot kg^{-1}} \times 30\ \mathrm{m}$$

$$= \mathbf{7.66\ MW}$$

$$\Omega_p = \frac{44.86(7.66 \times 10^6)^{1/2}}{1000^{1/2}(9.81 \times 30)^{5/4}}\ \mathrm{rad} = \mathbf{3.22\ rad}$$

13.11

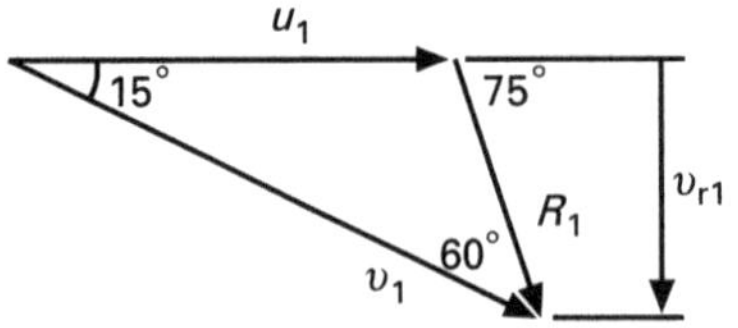

$$u_1 = 6.25\pi \times 0.76 \text{ m} \cdot \text{s}^{-1}$$
$$= 14.92 \text{ m} \cdot \text{s}^{-1}$$
$$v_1 = \frac{u_1}{\sin 60°} \sin 105°$$
$$= 16.64 \text{ m} \cdot \text{s}^{-1}$$
$$v_{r1} = 16.64 \sin 15° \text{ m} \cdot \text{s}^{-1}$$
$$= 4.31 \text{ m} \cdot \text{s}^{-1}$$
$$\therefore Q = 4.31 \times 0.2 \text{ m}^3 \cdot \text{s}^{-1}$$
$$= 0.862 \text{ m}^3 \cdot \text{s}^{-1}$$
$$v_{w1} = v_1 \cos 15° = 16.08 \text{ m} \cdot \text{s}^{-1}$$

$$\therefore \eta_{hyd} = \frac{u_1 v_{w1}}{gH} = \frac{14.92 \times 16.08}{9.81 \times 30} = \mathbf{81.5\%}$$

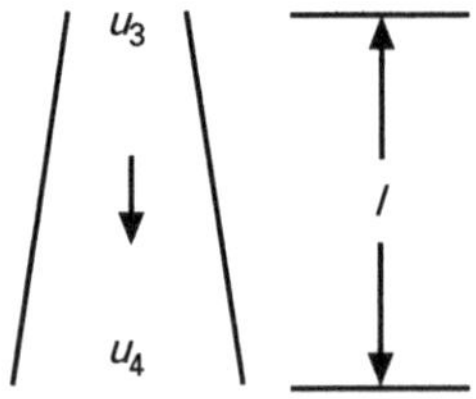

Steady-flow energy eqn:

$$H_3 + \frac{v_3^2}{2g} + l - 0.03Q^2 l = 10.33 \text{ m} + \frac{v_4^2}{2g} + 0$$

$$l(1 - 0.03Q^2) = (10.33 - 3.6) \text{ m} + \frac{1}{2g}\left(v_4^2 - v_3^2\right)$$
$$= 6.73 \text{ m} + \frac{Q^2}{2g}\frac{16}{\pi^2}\left(\frac{1}{d_4^4} - \frac{1}{d_3^4}\right)$$

$$0.978l = 6.73 + 0.0613\left(\frac{1}{d_4^4} - \frac{1}{0.450^4}\right)$$
$$= 5.234 + \frac{0.0613}{(0.450 + 2l \tan 8°)^4} \text{ [metre, second units]}$$

whence $l = \mathbf{5.36\ m}$

13.12 $$\Omega_p = \frac{52.3\left(15 \times 10^6\right)^{1/2}}{1000^{1/2}(9.81 \times 180)^{5/4}} \text{ rad} = 0.56 \text{ rad} = 0.56/2\pi \text{ rev}$$
$$= 0.089 \text{ rev}$$

$\therefore \sigma_c = 0.037$ say (Fig. 13.19)

$$z_{max} = \frac{p_a - p_v}{\varrho g} - \sigma_c H = (8.6 - 0.037 \times 180) \text{ m} = \mathbf{1.94\ m}$$

13.13 $H_{\text{man}} = (4 + 16.5)\text{ m} = 20.5\text{ m}$

$u_2 = 25\pi 0.25\text{ m}\cdot\text{s}^{-1} = 19.36\text{ m}\cdot\text{s}^{-1}$

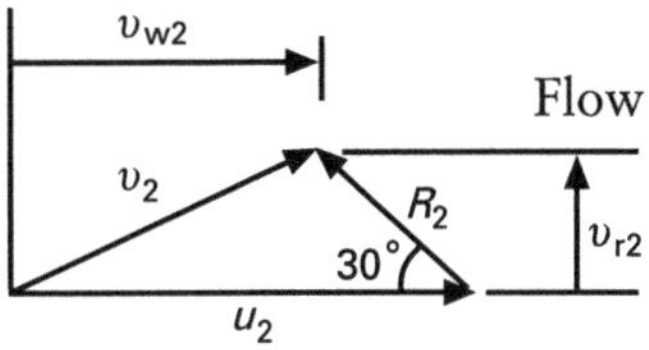

$$\text{Flow area} = (0.25\pi - 10 \times 0.005\,\text{cosec}\,30°)0.0125\text{ m}^2 = 0.00857\text{ m}^2$$

$$\therefore v_{r2} = \frac{0.026\text{ m}^3\cdot\text{s}^{-1}}{0.00857\text{ m}^2} = 3.035\text{ m}\cdot\text{s}^{-1}$$

$$v_{w2} = (19.63 - 3.035\cot 30°)\text{ m}\cdot\text{s}^{-1} = 14.38\text{ m}\cdot\text{s}^{-1}$$

$$\therefore \eta_m = \frac{9.81 \times 20.5}{19.63 \times 14.38} = \mathbf{0.712}$$

$$\text{Ideal head} = (19.63 \times 14.38/9.81)\text{ m} = 28.78\text{ m}$$

$$v_2^2 = v_{w2}^2 + v_{r2}^2 = (14.38^2 + 3.035^2)\text{ m}^2\cdot\text{s}^{-2} = 216.0\text{ m}^2\cdot\text{s}^{-2}$$

Steady-flow energy eqn between pump inlet and runner outlet:

$$\frac{p_0^*}{\varrho g} + \frac{v_0^2}{2g} + \frac{u_2 v_{w2}}{g} - \text{Runner loss} = \frac{p_2^*}{\varrho g} + \frac{v_2^2}{2g}$$

Then $$\frac{p_2^*}{\varrho g} + 0.5\frac{v_2^2}{2g} = \frac{p_3^*}{\varrho g}$$

$$\text{Hence Runner loss} = \frac{v_0^2}{2g} - 0.5\frac{v_2^2}{2g} + \frac{u_2 v_{w2}}{g} - \frac{p_3^* - p_0^*}{\varrho g}$$

$$v_0 = \frac{0.026\text{ m}^3\cdot\text{s}^{-1}}{(\pi/4)0.12^2\text{ m}^2} = 2.299\text{ m}\cdot\text{s}^{-1} \quad \therefore\ v_0^2/2g = 0.269\text{ m}$$

$$\therefore \text{Runner loss} = (0.269 - 5.50 + 28.78 - 20.5)\text{ m} = \mathbf{3.05\text{ m}}$$

13.14 $u_2 = 13.5\pi 0.75\text{ m}\cdot\text{s}^{-1} = 31.81\text{ m}\cdot\text{s}^{-1}$

$$v_{r2} = \frac{5.7\text{ m}^3\cdot\text{s}^{-1}}{2\pi 0.375 \times 0.125\text{ m}^2} = 19.35\text{ m}\cdot\text{s}^{-1}$$

$$v_{w2} = u_2 - v_{r2}\cot 70° = 24.76\text{ m}\cdot\text{s}^{-1}$$

$$R_2 = v_{r2}\,\text{cosec}\,70° = 20.60\text{ m}\cdot\text{s}^{-1}$$

$$v_2^2 = v_{w2}^2 + v_{r2}^2 = \left(24.76^2 + 19.35^2\right)\text{m}^2\cdot\text{s}^{-2} = 988\text{ m}^2\cdot\text{s}^{-2}$$

$$\text{Head of air} = \frac{1}{2g}\left(2u_2v_{w2} - 0.4R_2^2 - v_2^2 + 0.5v_2^2\right)$$
$$= \frac{1}{19.62}\left(2 \times 31.81 \times 24.76 - 0.4 \times 20.60^2 - \frac{1}{2} \times 988\right) \text{m}$$
$$= 46.5 \text{ m}$$

$$\text{Equivalent head of water} = 46.5 \text{ m} \times \frac{1.25}{1000}$$
$$= \mathbf{58.1 \text{ mm}}$$

$$\text{Power} = Q\varrho g\left(\frac{u_2v_{w2}}{g}\right) + \text{Mech. losses}$$
$$= (5.7 \times 1.25 \times 31.81 \times 24.76 + 220) \text{ W}$$
$$= \mathbf{5830 \text{ W}}$$

13.15

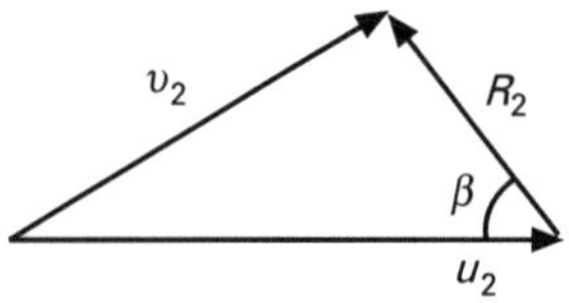

$$H = \frac{u_2v_{w2}}{g} - \frac{v_2^2}{2g};$$
$$v_{w2} = u_2 - v_{r2}\cot\beta$$
$$v_2^2 = v_{r2}^2 + (u_2 - v_{r2}\cot\beta)^2$$
$$\therefore H = \frac{1}{2g}\left\{2u_2(u_2 - v_{r2}\cot\beta) - v_{r2}^2 - u_2^2 + 2u_2v_{r2}\cot\beta - v_{r2}^2\cot^2\beta\right\}$$
$$= \frac{1}{2g}\left\{u_2^2 - v_{r2}^2\text{cosec}^2\beta\right\}$$

$$u_2 = 30\pi 0.5 \text{ m}\cdot\text{s}^{-1} = 47.1 \text{ m}\cdot\text{s}^{-1};$$
$$v_{r2} = \frac{4.5 \text{ m}^3\cdot\text{s}^{-1}}{\pi 0.5 \times 0.18 \text{ m}^2} = 15.92 \text{ m}\cdot\text{s}^{-1}$$
$$\therefore 0.1 \text{ m} \times \frac{1000}{1.23} = \frac{1}{19.62}(47.1^2 - 15.92^2\text{cosec}^2\beta) \text{ m}$$

whence $\beta = \mathbf{39.5^\circ}$ or 140.5°

[Reject second result as too extreme.]

13.16 Head to be supplied $= (2.4 + 19 + 68Q^2 + 650Q^2)$ m
$$= (21.4 + 718Q^2) \text{ m}$$

$$u_2 = 16.6\pi 0.35 \text{ m}\cdot\text{s}^{-1} = 18.25 \text{ m}\cdot\text{s}^{-1}$$

$$v_{r2} = \frac{Q}{0.95\pi D_2 B_2} = \frac{Q}{0.95\pi 0.35 \times 0.018}\ \mathrm{m \cdot s^{-1}} = 53.2Q\ \mathrm{m \cdot s^{-1}}$$

$$v_{w2} = (18.25 - 53.2Q \cot 35°)\ \mathrm{m \cdot s^{-1}} = (18.25 - 76.0Q)\ \mathrm{m \cdot s^{-1}}$$

$$\therefore (21.4 + 718Q^2)\ \mathrm{m} = \frac{u_2 v_{w2}}{g} - \frac{v_2^2}{2g} + 0.5\frac{v_2^2}{2g}$$

$$= \frac{1}{19.62}\Big[2 \times 18.25(18.25 - 76.0Q)$$

$$-0.5\left\{53.2^2Q^2 + (18.25 - 76.0Q)^2\right\}\Big]\ \mathrm{m}$$

$$\therefore 937Q^2 + 70.66Q - 4.07 = 0, \quad \text{whence} \quad Q = 0.03823$$

that is, Flow rate = **38.23 L · s^{-1}**

$$\eta_m = \frac{9.81\left\{21.4 + 718(0.03823)^2\right\}}{18.25(18.25 - 76.0 \times 0.03823)} = \mathbf{78.6\%}$$

13.17 Velocity in pipes $= \dfrac{0.045\ \mathrm{m^3 \cdot s^{-1}}}{(\pi/4)(0.15\ \mathrm{m})^2} = 2.546\ \mathrm{m \cdot s^{-1}}$

$$\therefore \text{Total losses} = \left(\frac{4 \times 0.006 \times 36}{0.15} + 2.4\right)\frac{2.546^2}{19.62}\ \mathrm{m} = 2.697\ \mathrm{m}$$

$\therefore$ Manometric head = (30 + 2.697) m = 32.7 m

$$0.465\ \mathrm{rad} = \frac{\omega(0.045\ \mathrm{m^3 \cdot s^{-1}})^{1/2}}{(9.81\ \mathrm{m \cdot s^{-2}} \times 32.7\ \mathrm{m})^{3/4}} \quad \therefore \omega = 166.2\ \mathrm{rad \cdot s^{-1}}$$

$$\text{Area of flow} = \pi D \times \frac{D}{10} \times 0.95 = 0.2985D^2$$

$$\therefore v_{r2} = \frac{0.045\ \mathrm{m^3 \cdot s^{-1}}}{0.2985D^2} = \frac{0.1508\ \mathrm{m^3 \cdot s^{-1}}}{D^2}$$

$$\therefore u_2 = 166.2D/2\ \mathrm{s^{-1}} = 83.1D\ \mathrm{s^{-1}}$$

$$\frac{gH}{v_2 v_{w2}} = 0.75 \quad \therefore v_{w2} = \frac{9.81 \times 32.7}{0.75 \times 83.1D}\ \mathrm{m^2 \cdot s^{-1}} = \frac{5.15}{D}\ \mathrm{m^2 \cdot s^{-1}}$$

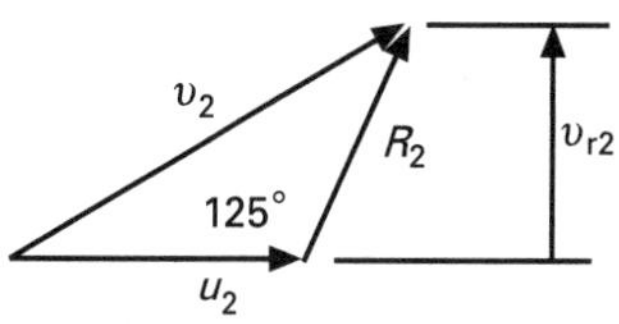

$$\tan 55° = \frac{v_{r2}}{v_{w2} - u_2}$$

$$= \frac{0.1508}{D_2\left[(5.15/D) - 83.1D\right]}$$

[metre, second units]

$$\therefore 5.15D - 83.1D^3 = 0.1056$$

As first approx. neglect RHS. Then $D = (5.15/83.1)^{1/2} = 0.249$

Next $D^3 = \dfrac{1}{83.1}(5.15 \times 0.249 - 0.1056) = 0.01416$ giving

$D = 0.242$ and so on.

Exact $D =$ **0.238 m, say 250 mm**

13.18 $u_2 = 24.2\pi 0.25\ \mathrm{m \cdot s^{-1}} = 19.01\ \mathrm{m \cdot s^{-1}}$

$$v_{r2} = \frac{0.03\ \mathrm{m^3 \cdot s^{-1}}}{17\,000 \times 10^{-6}\ \mathrm{m^2}} = 1.765\ \mathrm{m \cdot s^{-1}}$$

Apparent $v_{w2} = u_2 - v_{r2} \cot 32^\circ = 16.18\ \mathrm{m \cdot s^{-1}}$

$\therefore$ Actual $v_{w2} = 0.7 \times 16.18\ \mathrm{m \cdot s^{-1}} = 11.33\ \mathrm{m \cdot s^{-1}}$

$$\text{Velocity in discharge pipe} = \frac{0.03}{(\pi/4)0.125^2}\ \mathrm{m \cdot s^{-1}} = 2.445\ \mathrm{m \cdot s^{-1}}$$

$$\text{Velocity in suction pipe} = \frac{0.03}{(\pi/4)0.125^2}\ \mathrm{m \cdot s^{-1}} = 1.698\ \mathrm{m \cdot s^{-1}}$$

$$\text{Overall head} = \left\{4.5 + 13.3 + \frac{1}{19.62}(2.445^2 - 1.698^2)\right\}\ \mathrm{m}$$

$$= 17.96\ \mathrm{m}$$

$$\eta_{\text{overall}} = \frac{0.3 \times 1000 \times 9.81 \times 17.96}{7760} = \mathbf{68.1\%}$$

$$\eta_{\text{man}} = \frac{9.81(4.5 + 13.3)}{19.01 \times 11.33} = \mathbf{81.1\%}$$

13.19 (a) $\Omega_s = \dfrac{\omega(14)^{1/2}}{(9.81 \times 1.5)^{3/4}}\ [\mathrm{s}] = 0.498\omega\ [\mathrm{s}]$

n	40	39	38	pairs of poles
$\therefore N$	1.25	1.282	1.316	rev/s
ω	7.85	8.05	8.27	$\mathrm{rad \cdot s^{-1}}$
Ω_s	3.91	4.00	4.12	rad

Hence $\Omega_s = 4$ rad and pump speed $= 8.05\ \mathrm{rad \cdot s^{-1}}$

(b) $70\ \mathrm{kPa} \equiv \dfrac{70 \times 10^3}{800 \times 9.81}\ \mathrm{m} = 8.92$ m head

$$\Omega_s = \frac{\omega(0.0113)^{1/2}}{(9.81 \times 8.92)^{3/4}}\ [\mathrm{s}] = 0.003716\omega\ [\mathrm{s}]$$

Highest possible speed $= 50$ rev/s $= 314.2\ \mathrm{rad \cdot s^{-1}}$

Then $\Omega_s = 0.003716 \times 314.2 \text{ rad} = 1.17 \text{ rad}$, say **1.20 rad**

Alternatively use $\Omega_s = 0.60$ rad at 157 $\text{rad} \cdot \text{s}^{-1}$ but pump would then be more bulky.

(c) $gH = p/\varrho = \dfrac{5.5 \times 10^6}{1000} \text{ m}^2 \cdot \text{s}^{-2} = 5500 \text{ m}^2 \cdot \text{s}^{-2}$

At 50 rev/s, $\Omega_s = \dfrac{2\pi \times 50(0.00525)^{1/2}}{(5500)^{3/4}} \text{ rad} = 0.0356 \text{ rad}$

—far too low for range of pumps

∴ Use multi-stage pump, each stage with

$\Omega_s =$ **0.20 rad at 314.2 rad · s^{-1}**

$$0.20 \text{ rad} = \frac{2\pi\, 50(0.00525)^{1/2}}{(gH)^{3/4}} \text{ rad} \cdot \text{m}^{3/2} \cdot \text{s}^{-3/2}$$

$\therefore gH = 548 \text{ m}^2 \cdot \text{s}^{-2}$

∴ **10 stages required**

13.20 $\sigma_c = \dfrac{3.26 - 1800/(1000 \times 9.81)}{36.5} = \mathbf{0.0843}$

$$z_1 - z_2 = (0.750 - 0.622)13.56 \text{ m} = \frac{1800 - 830}{1000 \times 9.81} \text{ m} = \mathbf{1.637\ m}$$

13.21 By dimensional analysis $\dfrac{Q}{\omega D^3} = \phi\left(\dfrac{gH}{\omega^2 D^2}, \dfrac{\omega D^2}{\nu}\right)$

For prototype pump $1.15 \text{ rad} = \dfrac{\omega\sqrt{2}}{(9.81 \times 15)^{3/4}} \text{ s}$

whence $\omega = 34.37 \text{ rad} \cdot \text{s}^{-1}$

$$\text{For equal } \omega D^2/\nu, \quad \omega_{\text{model}} = \omega_p \left(\frac{D_p}{D_m}\right)^2 \frac{\nu_m}{\nu_p}$$

$$= \left(34.37 \text{ rad} \cdot \text{s}^{-1}\right) 4^2 \frac{1}{n}$$

$$= \frac{549.8}{n} \text{ rad} \cdot \text{s}^{-1} \text{ where } \quad 3 \leqslant n \leqslant 6$$

$\therefore$ **91.6 rad · s^{-1}** $\leqslant \omega_m \leqslant$ **183.2 rad · s^{-1}**

$$\text{For equal } gH/\omega^2D^2,\; H_m = H_p\left(\frac{\omega_m}{\omega_p}\right)^2\left(\frac{D_m}{D_p}\right)^2$$

$$= H_p\left(\frac{D_p}{D_m}\right)^4\left(\frac{v_m}{v_p}\right)^2\left(\frac{D_m}{D_p}\right)^2$$

$$= (15\text{ m})\left(\frac{D_p}{D_m}\right)^2\frac{1}{n^2} = \frac{240}{n^2}\text{ m}$$

$\therefore$ **6.67 m $\leqslant H_m \leqslant$ 26.67 m**

13.22 Input power $= T\omega_1 = C_T\varrho\omega_1^3D^5$

$$\therefore \text{Lost power} = C_T\varrho\omega_1^3D^5s$$

$$= 0.0014 \times 850\text{ kg}\cdot\text{m}^{-3}\left(16.67 \times 2\pi\text{ rad}\cdot\text{s}^{-1}\right)^3(0.5\text{ m})^5 0.03$$

$$= \mathbf{1282\ W}$$

13.23 From eqns 13.29 and 13.30,

$$\frac{1}{g}\left\{\left(\omega_1^2r_0^2 - \omega_1\omega_2r_i^2\right) - \left(\omega_1\omega_2r_0^2 - \omega_2^2r_i\right)\right\} = 4\frac{v^2}{2g}$$

that is, $(\omega_1r_0^2 - \omega_2r_i^2)(\omega_1 - \omega_2) = 2v^2$

Power transmitted $= Q\varrho$ (Work divided by mass in secondary)

$$= Av\varrho\left(\omega_1\omega_2r_0^2 - \omega_2^2r_i^2\right)$$

$$= A\varrho\left\{\frac{1}{2}\left(\omega_1r_0^2 - \omega_2r_i^2\right)(\omega_1 - \omega_2)\right\}^{1/2}(\omega_1\omega_2r_0^2 - \omega_2^2r_i^2)$$

$$= A\varrho\omega_2\left(\omega_1r_0^2 - \omega_2r_i\right)^{3/2}\left(\frac{\omega_1 - \omega_2}{2}\right)^{1/2}$$

$$\therefore 150 \times 10^3\text{ W} = 0.026\text{m}^2 \times 850\text{ kg}\cdot\text{m}^{-3} \times 40 \times 2\pi\text{ rad}\cdot\text{s}^{-1} \times 0.965\left(80\pi 0.19^2\text{ m}^2\cdot\text{s}^{-1} - 80\pi 0.965\text{ rad}\cdot\text{s}^{-1}r_1^2\right)^{3/2} \times \left\{\frac{80\pi(1 - 0.965)}{2}\right\}^{1/2}(\text{rad}\cdot\text{s}^{-1})^{1/2}$$

$$\therefore (0.19^2\text{ m}^2 - 0.965r_1^2)^{3/2} = 0.003349\text{ m}^3$$

Whence $r_i = 0.1192$ m $\quad\therefore d_i =$ **238.4 mm**